T0212396

Dual Mode Logic

Itamar Levi • Alexander Fish

Dual Mode Logic

A New Paradigm for Digital IC Design

 Springer

Itamar Levi
Faculty of Engineering
Bar-Ilan University
Ramat Gan, Israel

Alexander Fish
Faculty of Engineering
Bar-Ilan University
Ramat Gan, Israel

ISBN 978-3-030-40788-9   ISBN 978-3-030-40786-5   (eBook)
https://doi.org/10.1007/978-3-030-40786-5

This Springer imprint is published by the registered company Springer Nature Switzerland AG
The registered company address is: Gewerbestrasse 11, 6330 Cham, Switzerland

We are eternally grateful to our dear families who helped us be dual-mode *fathers and husbands throughout the long period we spent writing this book. Without your support, we would never have been able to bring it to fruition.*

Dear Anat, my beautiful wife, and my wonderful children, Arad and Negev, thank you for putting up with me (or without me), Itamar.

Dear Marianna, my beloved wife, and my beautiful children, Tom, Daniel and Yonatan, thank you for your support and constant reminders to work on the book whenever I had spare time, Alex.

Preface

Digital integrated circuits play an important role in the energy, performance, reliability, and cost targets of almost all electronic devices and silicon-based systems. Over the last four decades, conventional Static Complementary Metal Oxide Semiconductor (CMOS) Logic has become the dominant design methodology and is now widely used in the semiconductor industry. The advantages of CMOS design include strong "on" and "off" digital states, low leakage energy, high reliability, and simple physical implementation. These features have led to the standardization of CMOS, which has resulted in the development of CMOS-compatible EDA tools and CMOS-based digital libraries. However, the complementary nature of CMOS gates, which require 2N transistors for the implementation of an N-input gate, translates into the large area and capacitance of the gate and results in relatively high power consumption and poor performance. The energy–delay (E-D) tradeoff in CMOS is considerable, making it very challenging to optimize CMOS gates simultaneously for speed and energy efficiency. Although static CMOS Logic remains the most popular design approach, many attempts have been made to find a better alternative to achieve lower energy consumption, a smaller size, and better performance. However, no one solution has managed to tackle all three metrics successfully.

This book attempts to provide a comprehensive solution to this problem by presenting an innovative design paradigm for digital IC design, dubbed Dual Mode Logic (DML). DML logic gates can operate in two modes, each optimized for a different metric. DML designs allow on-the-fly switching between the operational modes at the gate, block, and system levels and provide a very high level of E-D optimization flexibility. We show that the DML paradigm makes it possible to implement digital circuits that dissipate less energy while simultaneously improving performance and reducing area. All these are achieved without a significant compromise in reliability. DML design methodology is shown to be compatible with existing standard electronic design automation (EDA) tools, thus providing the opportunity for a new generation of DML-based designs as a solid alternative to CMOS. This book covers all aspects of DML methodology, starting from the basic concept, through single gate optimization, the general module optimization, design tradeoffs, and new ways in which DML can be integrated into standard design flows. Scalability of DML gates is also discussed. The first part of this book (Chaps. 1, 2, 3, 4, and 5) presents the concepts behind basic DML operations and provides a detailed exploration of design tradeoffs and optimization methods. The second part (Chaps. 6, 7, 8, and 9) shows how DML modes of operation can be controlled on the fly from the architecture to the gate levels. The compatibility of DML with standard design flow is analyzed as well as the unique capabilities of DML in advanced technologies.

Researchers, engineers, and graduate students can all draw on this book to design and optimize advanced energy efficient, high performance digital ICs for a wide range of applications. DML can be used in conjunction with any other design style so designers can implement and integrate the DML solution with existing designs, thus significantly improving the power and performance of these designs.

This book is the result of extensive research, conducted by the editors along with other faculty members, postdoctoral fellows, Ph.D., and Master's students from Bar Ilan (BIU) and Ben Gurion (BGU) Universities, Israel, EPFL Switzerland as well as the University of Calabria, Italy. Sagi Fisher contributed to introduction of the DML concept in general, and subthreshold operation of DML gates in particular. Asaf Kaizerman was involved in low voltage DML operation, model analysis, and parameter extraction. Dr. Alexander Belenky was the co-developer of the logic effort design methodology for DML. Dr. Adam Teman, Viacheslav Yuzhaninov, and Lior Atias contributed to the development of the design flow, synthesis, and characterization methodology for DML. Prof. Shmuel Wimer and Amir Albek were closely involved in Dual Mode Square (DM^2) concept development. This concept is a key part of the DML control methodology at the architecture level. Dr. Ramiro Taco and Prof. Marco Lanuzza led the research on optimization of DML on FD-SOI process and contributed to the evaluation of DML in nanoscaled technologies. Netanel Shavit and Inbal Stanger worked on the implementation of DML in FinFet process.

The development of the DML was supported by the Israel Innovation Authority and the Israel Science Foundation. Specifically, Israel Innovation Authority (IIA) MAGNET programs have supported DML-related projects through the High-Performance (HiPer) consortium and the Generic Processor (GenPro) consortium. The content is based on more than 13 scientific journal articles and conference papers.

A special thanks goes to TSMC and STMicroelectronics for their support with test-chips design and fabrication throughout the research in various technology nodes and a variety of processes; and to Cadence and Synopsis for their help with tailored tools and adaptation of design stages for advanced research purposes.

Ramat-Gan, Israel   Itamar Levi
Ramat-Gan, Israel   Alexander Fish
November 2019

Contents

Acronyms

ASIC	Application specific integrated circuit
BGU	Ben Gurion University
BIU	Bar Ilan University
CLA	Carry look-ahead adder
CML	Current mode logic
CMOS	Complementary metal oxide semiconductor
CNFET	Carbon nanotube field effect transistor
DIBL	Drain-induced barrier lowering
DML	Dual mode logic
DPL	Double pass transistor logic
DTMOS	Dynamic threshold CMOS
DVFS	Dynamic voltage–frequency scaling
EDA	Electronic design automation
EPFL	École Polytechnique Fédérale de Lausanne
FD-SOI	Fully depleted silicon on insulator
FET	Field effect transistor
GDI	Gate diffusion input
GIDL	Gate-induced drain leakage
GTL	Gate-level netlist
HDL	High description language
IC	Integrated circuit
IVC	Input vector control
LE	Logical effort
MAC	Multiplier and accumulator
MCML	MOS current mode logic
MDP	Minimum delay point
MEP	Minimum energy point
MOS	Metal oxide semiconductor
MOSFET	Metal oxide semiconductor field effect transistor
MRAM	Memristive RAM
MTJ	Magnetic tunnel junction

PDK	Process design kit
PTL	Pass transistor logic
RBB	Reversed body bias
RRAM	Resistivity RAM
SABL	Sense amplifier-based logic
SBB	Self-body-bias
SCL	Source coupled logic
SDF,STDF	Standard design flow
SNM	Static noise margin
SOI	Silicon on insulator
VLSI	Very large-scale integration
VTC	Voltage transfer characteristic

Chapter 1
Introduction

This chapter introduces the basic concepts and methodologies behind digital design. We start with current practices and the layout limitations of standard design methodologies. We then survey different alternatives for digital design (logic families) that can be implemented with standard CMOS processes. We discuss the tradeoffs and paradigms of energy and delay in digital designs. These lay the groundwork and provide the reader with the basic concepts. We end with a presentation of the general outline of this book.

1.1 Energy-Efficient and High-Performance Digital Design Limitations

Today's rapid advances in technology and the expansion of mobile applications have made energy consumption, which places one of the fundamental limits on both high-performance microprocessors and low-to-medium performance portable systems, a crucial issue in very large-scale integration (VLSI) digital design [1–10]. In high-performance systems, energy and peak power curtail further increases in performance and circuit density, because of the difficulties inherent to conveying power to circuits and removing the heat they generate. Correlatively, the integration of circuits with different workloads and activity profiles results in the formation of hot spots and temperature gradients over the die. This can impact long-term reliability and complicate system verification [11], thus turning temperature monitoring into a major component of design [12]. In portable battery-operated devices such as cellular phones, bio-medical instruments, sensor networks, etc., energy consumption is critical since it determines the lifetime of the battery (for non-rechargeables) or the time between recharges. It also affects packaging, cost, and weight.

Many architectures and techniques have been researched, analyzed and fabricated for power reduction and energy minimization of combinational circuits

© Springer Nature Switzerland AG 2021
I. Levi, A. Fish, *Dual Mode Logic*, https://doi.org/10.1007/978-3-030-40786-5_1

[1, 13–17]. In general, power reduction methodologies can be implemented at different levels of design abstraction from the system, algorithm, architecture, gate, and circuit to the level of the technology itself. At the algorithm level, for example, methods for simplifying the logic involved in computation and coding for smaller Hamming distances have been developed [18–20]. Examples of energy reduction design approaches at the system level include the ubiquitous power-saving modes, dynamic voltage scaling [21–23], clock gating [24–27], and leakage power reduction management [28, 29]. Other examples of highly efficient system level solutions include the Razor-2 processor [15] which takes process voltage temperature drifts into account, the Razor-free architecture [16], and low-power wake-up modes [30, 31]. The sleepy-stack/keeper approach to reducing leakage power [29], reduced swing techniques by special circuit families such as current mode logic (CML) [32] and reduced swing circuits [33], advanced subthreshold device sizing and balancing optimization techniques such as [34], and the use of high-cost functions for power hungry gates while optimizing circuits are all part of the panoply of current techniques to reduce power at the gate level. A variety of design techniques, such as input vector control (IVC) [35–37], the reverse body bias (RBB) technique [38–41], self-body bias (SBB) and unique control mechanisms [17, 42], dynamic threshold CMOS (DTMOS) [43, 44], and numerous others, can be implemented at the circuit and transistor levels to reduce both dynamic power and static power [3, 45–48]. Finally, at the technology level, fabrication technology that can operate under supply voltage reduction can substantially reduce both dynamic and leakage power [8, 49, 50]. Novel devices and integration methodologies have become increasingly available for mass production (e.g., RRAM, CNFET, and 3D integration [51] and MRAM devices [52]); in turn, these emerging technologies suggest that significant advances can be made in the near future.

During the process of standard design flow (SDF), the logical representation of integrated circuits (ICs) (in high description language (HDL)) is synthesized to a library of standard digital cells. These logical cells are implemented according to a well-optimized layout structure that also adheres to the device's dimensions and the physical guidelines of the digital library provider. Digital libraries have different flavors of gates. Most digital circuits are implemented using standard CMOS logic. Static CMOS logic has been the most popular design approach for the past 30 years or so. Many attempts have been made to come up with a better alternative logic family that would achieve lower power dissipation, take up less area, and yield higher performance. Early on, the pass transistor logic (PTL) was hailed as a promising alternative to static CMOS logic [45, 53–58]. Unfortunately, the leakage of PTL implementations of monotonic gates was shown to be much higher than that of CMOS implementations. Some PTL techniques, such as double PTL (DPL), were tested to solve these and other problems [56, 58, 59]. However, most solutions resulted in an increased transistor count and area, a large number of required buffers, and a degradation in signal integrity. Recently, a variety of interesting emerging logic families have been described that usually combine transistors with memristor devices [60, 61], solely on a magnetic tunnel junction [62], to implement logic gates. However, these families are not widely used since the fabrication process of memristors for this purpose is still not commercially available as of this writing.

Dynamic logic [63–65] has certain hallmarks of an efficient alternative for high-performance operations. Computation using dynamic gates operates in two phases: precharge and evaluation. The advantages of dynamic logic include the high driving strength of its evaluation network and its much lower transistor count. On the other hand, dynamic logic has a number of significant drawbacks such as charge leakages, charge sharing, signal integrity and restoration issues, high dynamic power consumption, and susceptibility to glitches with no data recovery. These sensitivities are intensified with process scale-down, supply voltage reduction, and increased process variations.

In a given standard cell library, the same gate can be designed in several flavors that are optimized for different objectives. While utilizing a digital logic family such as CMOS, the traditional paradigm assumes that high performance comes at the expense of energy efficiency; that is, one can design low-energy cells that operate at low frequencies or high-performance cells that consume higher energy. Research and optimizations to disentangle this paradigm (i.e., getting the best of both worlds) are typically carried out at the algorithm or architectural levels of the design by operations in different modes, dynamic voltage, and frequency scaling (DVFS), in different power domains, algorithmic optimizations, etc. This objective is more difficult to achieve at lower abstraction levels such as the gate and transistor levels. When utilizing standard CMOS gates, the optimization space at the gate level is very small. This is primarily because the low-energy flavors of cells (gates) are typically bounded by the amount of energy reduction they can provide as compared to the high-performance flavors for the same supply voltage and frequency.

In traditional sequential designs, performance is dominated by the circuit's most critical (slowest) path, whereas energy consumption is basically the sum of all the design consumers (i.e., all the *gates/cells* in the design). The typical optimization strategy implemented by automated tools targeted by the specification to achieve both low-energy and high-energy performances is to find the design's most critical paths and to assign high-performance gates to them (gates consuming relatively high energy) while the rest of the paths (that are not timing-critical) are left to remain in a fairly low-performance mode that consumes little energy.

1.2 Introduction to the Design of Digital Logic Families

Numerous circuit styles can implement a given logic function. These styles are called logic families. Every logic family inherently has its own advantages and shortcomings, and the designer's choice of an appropriate family depends on the application and its specifications such as energy, performance, area, temperature, reliability, etc. The most common logic styles are complementary metal oxide semi-conductor (CMOS) [45–47, 53], pseudo-nMOS logic and pass transistor logic (PTL) [56–59], gate diffusion input (GDI) [7, 66–69], dynamic logic (i.e., Domino/CMOS

NORA, etc.) [63–65], current mode logic (CML) [32, 33], and sense amplifier based logic (SABL) [70, 71]. This range of styles has shrunk over the last 30 or so years to the handful of popular logic families found today (mainly CMOS for noncustom designs). However, as discussed above, new logic families are constantly being proposed (e.g., utilizing memristors [60, 61] or RRAM-based [72] and magnetic tunnel junction (MTJ) [62]). However, to date (and for the near future) process design kits (PDKs) which contain such devices are not likely to become available commercially, i.e., there are no fully characterized or tested libraries for high-yield IC fabrication. It is worth noting that some proposals involve building logic devices that decode ternary logic levels (e.g., [73]). Clearly, such advances should go hand in hand with optimization algorithms at higher abstraction levels such as synthesizers, partitioning algorithms, etc. Thus, current modern designs consist of a limited selection of well-known and explored logic families for automate or semi-automate design flow. In the following subsections, we present the key features of the most common logic families. We primarily address logic styles that serve as the basic building blocks for the DML family, which is the topic of this book.

1.2.1 Complementary Metal Oxide Semiconductor (CMOS)

The most common design logic family today is CMOS. It is based on the use of complementary MOS transistors to perform logic functions (see Fig. 1.1) with a very small consumption of static current. CMOS gates are based on the fundamental inverter circuit which consists of two transistors. Both transistors are MOSFETs, with one n-channel whose source is grounded, and one p-channel whose source is connected to the V_{DD}. Their gates are connected to form the input, and their drains are connected to form the output. The two MOSFETs are designed to have matching characteristics and thus are complementary. When off, their resistance is very high

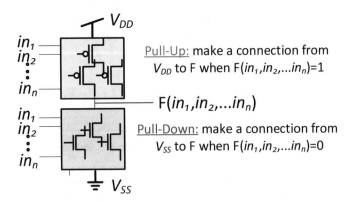

Fig. 1.1 Standard CMOS logic. Reproduced from [74]

(theoretically infinite), but when on, their channel resistance is low. Since the gate is essentially an oxide isolated circuit, it draws no current (except small gate leakage) in the steady state, and the output voltage is equal to one of the strong power supply voltages, depending on which transistor is conducting [74].

The advantages of conventional CMOS design methodology are well-known, so they are only partially detailed here. Its key features include strong on and off states, rail-to-rail logic levels, and until the advent of recent processes, very low static power consumption. The main drawbacks of CMOS consist of its large number of transistors (twice the number of inputs), which reflects the very large input and output capacitances responsible for increasing the delay. In most advanced nanoscale processes where the feature size is scaled to less than 65 nm, the static leakage current increases radically as a result of the increment in the subthreshold slope factor [1]. This problem constitutes a significant obstacle to low-voltage designs. The increased leakage results in a decreased on/off current ratio and thus increased delay, especially under low-voltage operation. The decreased on/off ratio is a recipe for high-probability failure mechanisms, especially when there are global process variations for high-frequency applications if these are not designed correctly [75, 76]. More typical sensitivities come into play with scaled process nodes such as gate-induced drain leakage (GIDL), drain-induced barrier lowering (DIBL) [76, 77], punch through [78], gate tunneling [79], etc. Some of these issues have been partially resolved by FinFET technology. CMOS performance in low-voltage regimes is discussed in the next subsection.

Threshold voltage needs to scale with frequency, current, and future CMOS technology generations. However, the techniques used to decrease the threshold voltage increase its variability and leakage power. In CMOS, gates are generally designed to have a stacked network and a parallel network whose leakage can be substantial. Thus, summing large leakage currents which are intensified by V_T variations makes the CMOS design particularly vulnerable to process variations in nanometer CMOS. V_T sensitivity is caused by random dopant fluctuations, line edge roughness, oxide thickness variations, etc. These effects combine to yield an exponential change in the ON or leakage (OFF) currents under low-voltage operation [80]. The utilization of low-voltage CMOS gates is associated with a number of challenges [80], including low worst-case frequency (which is determined by the design's most critical path), relatively large leakage currents in modern processes, as well as large areas and capacitances. All these make low-voltage CMOS design impractical in many applications where performance is important, and shunt CMOS low-voltage designs to applications such as medical devices [81, 82] and portable applications with very low performance requirements [6, 83].

Unlike the complementary CMOS which implements a symmetric voltage transfer characteristic (VTC) and has robust high and low noise margins, dynamic logic [84–86] was traditionally considered as the solution to counterbalance the symmetric nature of CMOS to provide high performance. Naturally this comes at the cost of greater energy consumption, as discussed next.

1.2.2   Dynamic Logic

Dynamic logic [84–86] was developed for high-performance digital circuits. It can be described as a technique to operate specially designed logic gates in two different phases: the precharge phase where the dynamic node is charged to logic level "1", and the evaluation phase which may lead to a discharge of the dynamic output node. The two phases are synchronized using a CLK signal. Figure 1.2 presents the basic scheme of a dynamic gate:

Unlike static families (constant low resistive connections to a constant supply of V_{DD} or GND, e.g., CMOS), a dynamic design has very high performance, has reduced area utilization, and inherently eliminates short circuit currents. Static gates can evolve to *high* or *low* logic levels with equal probability; for this reason, they are sized for equal high-to-low ($T_{P_{HL}}$) and low-to-high ($T_{P_{LH}}$) times. A dynamic gate has only one transition during evaluation, and each output can only change once (in contrast to CMOS where sporadic changes may occur). Therefore, several chargings of the same network do not take place in dynamic operation and the gates can be sized to optimize only one transition.

Dynamic logic presents a number of design challenges, especially in modern nanoscaled technologies. These challenges are associated with issues such as increased leakage currents, charge sharing, crosstalk sensitivity, signal integrity and restoration, high dynamic power consumption (at each cycle, even with clock gating), susceptibility to glitches with no data replenishment, and the increased complexity of dynamic design clock distribution and control. These sensitivities and challenges are intensified by the device dimensions, supply voltage scaling, process variations, and temperature fluctuations. Worse, there are no standard design flow tools or libraries for design with dynamic families (unlike CMOS). These challenges have significantly undermined the popularity of dynamic logic in the last 10 years. In fact, dynamic logic is almost never used in modern state-of-the-art

Fig. 1.2 Basic scheme of a dynamic gate. Reproduced from [85]

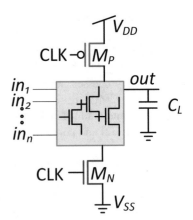

designs. However, as will be shown in the next sections on DML, the basic concepts of dynamic logic can be utilized in conjunction with other design styles to achieve fast and reliable operation.

1.2.2.1 The Cascading Challenge and Dynamic Logic Topologies

The cascading of several basic dynamic gates using the same clock signal (CLK) leads to a built-in race condition. An illustration of this race condition, depicted by two cascaded n-type dynamic inverters, is shown in Fig. 1.3.

During the precharge phase (i.e., CLK = "0"), the outputs of both inverters are precharged to V_{DD}. On the rising edge of the clock, output Out_1 starts to discharge. The second output needs to remain in the precharged state of V_{DD} because its expected value is "1" (Out_1 transitions to "0" during evaluation). However, there is a finite propagation delay for the input to discharge Out_1 to GND during which the second output also discharges. The conducting path is between Out_2 and GND, and precious charge is lost at Out_2 until Out_1 reaches V_{TH_n}. The conducting path is only disabled once Out_1 reaches V_{TH_n}. This leaves Out_2 at an intermediate voltage level and the correct level will not be recharged. Clearly, dynamic gates rely on capacitive storage unlike static gates such as CMOS. The charge loss leads to lesser noise margins and potential malfunctioning in all dynamic families and topologies. To remedy this, and achieve a correct-by-design implementation, two major dynamic design families were suggested:

- Domino logic
- NORA or np-CMOS logic

The domino logic topology ensures "0" in the gates' output nodes immediately after the precharge period. The cascading problem is resolved by adding an inverter after each dynamic gate that cuts off the next stage pull-down network during the

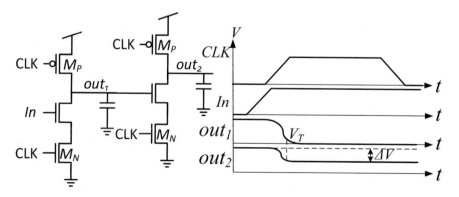

Fig. 1.3 Cascade of dynamic n-type blocks. Reproduced from [85]

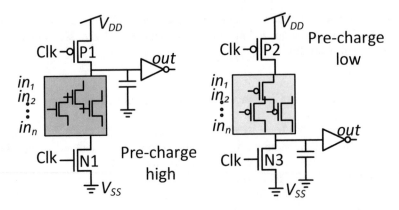

Fig. 1.4 Basic domino logic gate. Reproduced from [86]

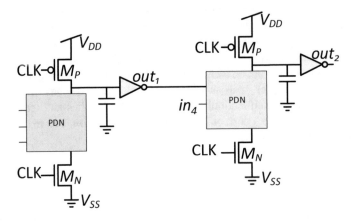

Fig. 1.5 Domino dynamic logic topology, cascaded gates

precharge period. Figure 1.4 outlines the general structure of a basic domino gate. Figure 1.5 provides an example of a cascade connection of precharge high gates.

The np-CMOS (alternatively, NORA) topology (Fig. 1.6) was introduced to improve performance and uses all the evaluation time for logic computation. This method constitutes an alternative to cascading dynamic logic by using two flavors (n-block and p-block types) of dynamic logic. In a p-block logic gate, $pMOS$ devices are used to build a PUN, including a $pMOS$ evaluation transistor. The $nMOS$ is a predischarge transistor that drives the output flow during precharge. The output makes a conditional "0" $\rightarrow$ "1" transition during evaluation depending on the p-block gates. The n-block gates are controlled by CLK, and the p-block gates are controlled by $\overline{CLK}$. n-block gates can push the p-block gates directly and vice versa. During the precharge phase (CLK = "0"), the output of the n-block gates, Out_1, is charged to V_{DD}, while the output of the p-block discharges to the ground ("0"). During evaluation, the output of the n-block gate can only switch

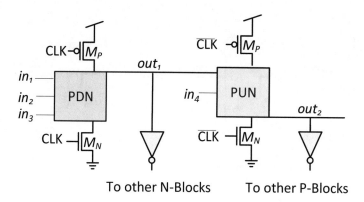

Fig. 1.6 np-CMOS dynamic logic topology

through a "1" → "0" transition that conditionally turns on some transistors in the p-block. This ensures that no false glitches take place. One of the shortcomings of this logic is that half of the total number of transistors are $pMOS$s. They are larger and slower (given their low driving strength) and have a lesser ability to push current in comparison to $nMOS$s (related to mobility differences). Therefore, under optimization that targets propagation delays, the $pMOS$s will consume a large area. However, in state-of-the-art processes such as FinFET, the difference between the driving strength of $pMOS$ and $nMOS$ is negligible.

1.2.2.2 A Footer Implementation

n-type dynamic gate inputs are low during precharge, so the designer may find it tempting to eliminate the evaluation transistor (footer). Clearly, elimination of the footer leads to a reduction in the sizes of the stacked devices in the evaluation network, less tree distribution efforts and load, and an increase in the pull-down drive strength. All these result in a better evaluation delay. Unfortunately, this change also prolongs the precharge period. In particular, the precharge will now ripple through the logic network, whereas originally it ran as a parallel operation for all gates in a chain at the same time. An example of a dynamic chain without footer transistors appears in Fig. 1.7.

An optional mitigation of the precharge ripple effect consists of the insertion of a logic stage with a footer every few stages. This is considered to be a reasonable compromise between the fast evaluation phase on one hand and an acceptable precharge phase duration on the other. Another critical negative effect of a rippling precharge is the extra power dissipation that takes place when both the pull-up and pull-down devices are on. To eliminate these short circuit currents, designers aim to delay the arrival of the clock between the dynamic gates, as shown in Fig. 1.8.

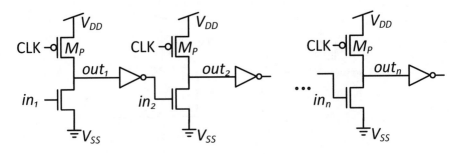

Fig. 1.7 Dynamic chain without footer transistors

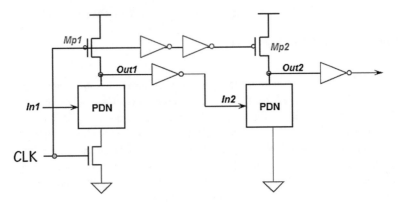

Fig. 1.8 Delay insertion for static power consumption mitigation

Thus, a footer constitutes a tradeoff between the evaluation phase duration and the subsequent precharge phase. Both periods form the operation cycle of a circuit, but typically the evaluation is much longer and the precharge phase can be concealed with multi-phasing clocks [85].

1.2.2.3 Low-Voltage Dynamic Logic

The rationale for operating dynamic logic in the subthreshold and near-threshold regions was to enhance performance as compared to low-voltage CMOS while still dissipating significantly less energy than in the super-threshold region. Attempts to use low-voltage dynamic logic were made in [84] where the conventional domino logic was implemented with an operational voltage below the transistor's threshold voltage. Although these efforts produced very interesting results, they are not practical for modern commercial applications because of their high sensitivity to process variations in advanced technologies. The use of dynamic logic in recent versions was also abandoned because of its increased control and clock complexity. In addition, domino circuits have been shown to exhibit poor tolerance to device subthreshold leakage [87]. Another attempt to modify standard dynamic logic to

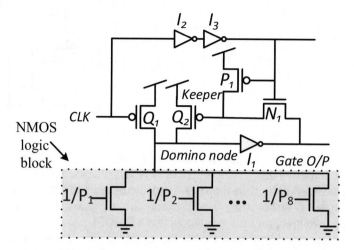

Fig. 1.9 An eight-input HS-domino OR gate. Reproduced from [63]

allow reliable $sub - V_{TH}$ operation was reported in [63], as shown in Fig. 1.9. Despite the fact that this system demonstrated high performance, it still suffered from high design complexity and area overhead.

Thus overall, dynamic logic operation with low voltage supply is daunting and complex and presents high sensitivities to process variations. Typically, it functions best in subthreshold and near-threshold designs if major modifications and extreme design efforts are implemented.

1.2.3 Other Design Styles in Standard CMOS Technology

This overview of existing logic families began with CMOS and dynamic logic families since they are the basic building blocks of a DML gate (as discussed in the next subsection). However, many other design styles have been suggested over the years in addition to CMOS and dynamic logics. In this section we overview a few of these alternative logic families.

1.2.3.1 Pass Transistor Logic (PTL)

In contrast to CMOS, which only addresses the primary inputs that drive the gate terminals of MOSFETs, pass transistor logic allows the primary inputs to drive source or drain terminals as well. The key advantage of this family is that a one pass transistor network (either $nMOS$ or $pMOS$) is sufficient to perform a logic computation, which results in a reduced transistor count and smaller input and output loads, in particular when using $nMOS$-based PTL cells. Similarly, the

Fig. 1.10 PTL AND gate

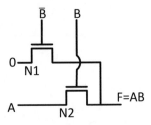

advantages of PTL have to do with its low transistor count, small loads, and, in some cases, improved delay. An example of an AND PTL gate appears in Fig. 1.10. However, the threshold voltage drop (V_T-drop) through the $nMOS$ transistors while passing a logic "1" makes swing (level) restoration at the gate outputs mandatory to avoid static currents at the next gate output (say a CMOS).

To decouple gate inputs and outputs and to provide acceptable output driving capabilities, inverters are usually attached to the gate outputs. Special design efforts must be made because in PTL, MOS networks are connected to the gate inputs rather than to constant power lines. If two networks are open ("On") at the same time while one drives GND and the other drives V_{DD}, it will lead to short circuit currents and high power consumption. This forbidden state is often called a "sneak path" and entails several logic restrictions when constructing PTL networks. Each PTL design must have a multiplexer general structure to avoid these "sneak paths." This issue is very restrictive and is one of the key obstacles to automate tools for PTL. For these reasons, certain restrictions need to be enforced when designing with PTL, i.e., both networks require the addition of a swing restoration circuit and an additional buffer. Furthermore, some PTL circuits are ratioed, which means that they are very sensitive to the sizing of their transistors. Hence, numerous sizing restrictions are required to preserve their functionality. The swing restoration circuitry (also called a keeper) is also size dependent and may not fully mitigate the static power. Usually, PTL circuits require an increased design effort (mostly custom and not automate design). While some PTL circuits are implemented in super-threshold designs, mainly for area reduction, they are sensitive to voltage scaling [57]. Under low voltage, PTL gates have reduced noise margins and experience reliability issues with low yield.

1.2.3.2   Gate Diffusion Input (GDI)

This logic family was presented by A. Morgenstein et al., in 2002 [66, 67]. This family combines notions deriving from the physical structures of CMOS and PTL gates. It is based on the use of a single simple cell, as shown in Fig. 1.11a. At first glance, the basic cell is reminiscent of the standard CMOS inverter. However, the GDI cell contains three inputs: G (the common gate input of both the $nMOS$ and the $pMOS$), P (the input to the source/drain of the $pMOS$), and N (the input to the source/drain of the $nMOS$). This simple cell can effectively implement a

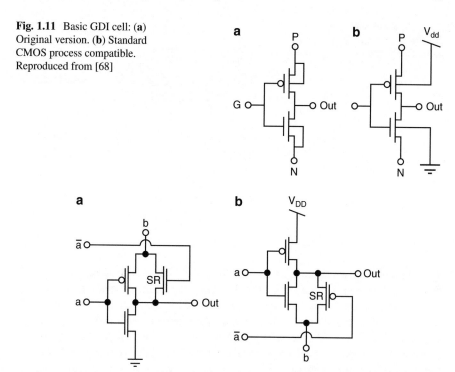

Fig. 1.11 Basic GDI cell: (**a**) Original version. (**b**) Standard CMOS process compatible. Reproduced from [68]

Fig. 1.12 (**a**) F1 with SR transistor. (**b**) F2 with SR transistor. Reproduced from [68, 88]

large range of Boolean functions, most of which are very complex in static CMOS. The GDI approach was originally developed for fabrication in silicon on insulator (SOI) and twin-well CMOS processes. The GDI cell shown in Fig. 1.11b [68] is a modification of the basic cell depicted in Fig. 1.11a and is also compatible with any standard CMOS fabrication process. In the past, various combinational and sequential circuits, such as adders, multipliers, comparators, flip-flops, and counters have been implemented in processes down to 65 nm and can exhibit a power reduction of up to 40%.

The main disadvantage of GDI is its lack of a full swing in some logic functions and its finite input resistance. Recently, a solution using a self-restoring (SR) transistor to regain the swing (two versions are shown in Fig. 1.12) [68, 88] was suggested. However, these gates are beyond the scope of automate design flows and have not been tested under low supply voltages.

1.2.3.3   Source-Coupled Logic (SCL) or Current Mode Logic (CML)

The general structure of source-coupled logic (SCL) (or current mode logic, CML) family gates [32, 33] was first implemented by bipolar transistors and extended to

Fig. 1.13 A conventional SCL-based inverter/buffer. Reproduced from [90]

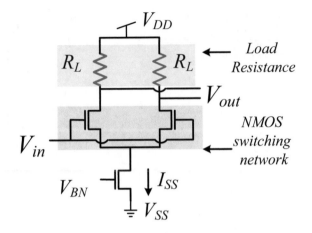

applications with MOS transistors. This logic, dubbed MOS current mode logic (MCML), is a useful logic style for the implementation of high-speed circuits. These circuits operate with a constant bias current for each gate and are appropriate for markets requiring accurate high-speed mixed signal applications. The general structure of this type of gate is depicted in Fig. 1.13. Typically, SCL circuits are differential and need a special analog layout periphery. These circuits are primarily for mixed-signal applications where dI/dt is of major importance (such as D/A or A/D converters). Due to the almost constant current and their inherent small voltage swing, they are highly immune to crosstalk noise incurred by switching current spikes (unlike CMOS) and are also advantageous for cryptographic circuits. Whereas MCML dissipates constant static power (the main power dissipation source of MCML), they require less dynamic power than conventional logic because of their smaller output swings, which are controlled and hence not sensitive to voltage scaling. The reduced output swing for each differential node is induced and controlled by the $pMOS$ resistor and the saturated constant current $nMOS$. These are designed for a constant reduced swing for the whole circuit. The reduced swing makes switching rapid, which is the main performance boost in these circuits. M. Alioto et al. [89] investigated the fit of this logic to the subthreshold regime. They showed that the total energy consumed by a MCML circuit in subthreshold (static power) could be less than the total energy (switching plus static) of a CMOS circuit. In [90] the values of a 10 pA constant bias current were listed for a subthreshold MCML circuit. Although these results are striking and promising, many aspects of MCML implementation in large scale and advanced nodes still require study. In addition, the compatibility of the MCML to standard design flow and tools needs to be examined.

1.3   Energy–Delay (E–D) Tradeoff Paradigms

For years, energy efficiency and performance have been the main metrics to assess digital electronic systems. Traditionally, energy efficiency comes at the expense of high performance and vice versa. R. G. Dreslinski et al., presented the general tradeoff of energy and performance as a function of the supply voltage (V_{DD}) [91]. Figures 1.14 and 1.15 present well-known illustrations of the tradeoffs between energy per operation and speed for different supply voltages.

The traditional designs that dominate today's market use supply voltages in the range of 0.9 V–1.8 V to operate their digital and analog circuits. In these designs all the "on" transistors operate in what is known as the super-V_T region (see Fig. 1.14), far above the switching threshold of a transistor. In this region the "Ion" current of the transistor used for switching digital gates from one state to another is very strong, which leads to a ratio of many orders of magnitude between the "Ion" and the "Ioff" (parasitic leakage) currents. This makes the super-V_T operation very fast

Fig. 1.14 Energy and performance tradeoff for different supply voltages. Reproduced from [91]

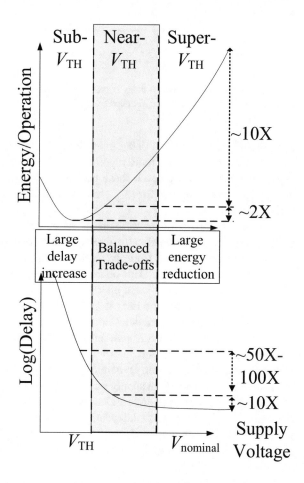

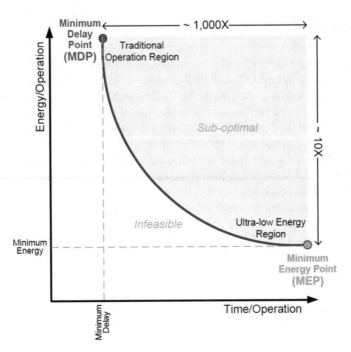

Fig. 1.15 Pareto-optimal energy–delay curve showing the minimum energy point (MEP) and the minimum delay point (MDP). Reproduced from [7]

and reliable. The minimum delay point (MDP), which indicates the best possible performance of a given circuit topology, is obtained in the super-V_T region, as depicted in Fig. 1.15. However, these traditional designs consume vast amounts of energy and are not suitable for many modern applications where energy dissipation is the main concern.

Low-voltage operation in the sub-V_T or near-V_T regions has been shown to be the ideal way to dramatically reduce energy dissipation. In sub-V_T designs, all transistors are operated from the supply voltage, which is below the transistor switching threshold voltage. This approach, which was put forward as early as the 1960s, is radically different since operation in the subthreshold regions exploits the parasitic leakage current and uses it as its primary operation current. This means that according to the classical definition, most of the transistors are in the "off" state even during their operation and switching. Subthreshold operation can substantially reduce both leakage and switching (dynamic) energy dissipation and thus results in minimum energy dissipation. Dynamic power is greatly reduced, primarily due to the quadratic dependency on supply voltage and the complete elimination of certain components. Similarly, the static leakage is also much lower since it also depends on the supply voltage (exponential dependency at low voltages). The minimum energy point (MEP) is usually situated in the subthreshold region, as shown in Figs. 1.14 and 1.15. Lowering the supply voltage below the MEP causes an increase in energy consumption (see Fig. 1.14). Below the MEP, the circuit delay is exponentially larger (an exponential increase with supply voltage reduction). In turn, this makes the leakage currents of sub-V_T devices start to dominate the energy per operation.

Although maximum energy efficiency is achieved in the sub-V_T region because the subthreshold currents are much weaker than standard "super-threshold" currents, the time needed to change a digital gate state is significantly longer, which limits the operation frequency of the circuit considerably. At the MEP, the delay is at least three orders of magnitude greater than at the MDP, which consumes an order of magnitude more energy. The energy–delay (E–D) curve is practically flat around the MEP, such that significant performance improvement can be achieved by a slight increase in the supply voltage above the MEP and by moving into the near-V_T region (see Fig. 1.15). In near-V_T (NT) designs, all the transistors are operated between the weak and moderate inversion regions. Since operation in the NT region presents a good tradeoff in terms of energy performance and leads to significant energy reduction with only a moderate drop in performance (as compared to the super-threshold region), it has become very attractive for many modern applications.

At first glance this description suggests that a simple reduction in the power supply voltage of traditional circuits could provide reliable $sub - /near$-threshold operation. Unfortunately, this turns out to be false, since the power supply reduction is accompanied by a number of problems and significant challenges. The low voltage associated with frequency reduction is not suitable for all modes of operation so that an adaptive voltage control mechanism may be required. Lower supply voltages also mean lower noise margins, a reduced yield, and increased vulnerability to process variations and temperature fluctuations. The characteristics of semiconductor behavior in the $sub - /near$-threshold are not well represented by standard transistor models and differ from those in the super-threshold region, resulting in variations in device sizing and ratio optimizations. Although some logic families such as CMOS are known to be fully operational in all voltage regions, many logic families including the ratioed-logic and dynamic families have been shown to be inapplicable in low-voltage regions. By contrast, the dual-mode logic (DML) discussed in this book is fully functional at all voltage regions. Crucially, the DML's capability to switch between different modes of operation (static or dynamic) enables it to achieve energy efficient operation under a wide range of supply voltages, while exhibiting a significant improvement in performance as compared to conventional CMOS logic.

1.4 Book Outline

After a general introduction to logic families and energy–delay tradeoffs in digital design, Chap. 2 introduces the reader to the basics of the DML logic family. This chapter consists of a description of the DML paradigm that depicts the transistor level architecture of basic DML gates and the principles of their operation. The optimization of DML gates is presented in Chap. 3. A dedicated logical effort

(LE) approach was chosen to illustrate the optimization of sizing and topology of DML gates. First, the optimization is shown for a simple case of a chain of inverters, and un-approximated, approximated, and semi-approximated methods are discussed to characterize the tradeoff between the precision and complexity of the model. Then, a general DML–LE method is introduced to optimize more complex gates and branches. The methodology is evaluated at the end of the chapter. Chapter 4 details the operation of DML circuits at low voltages. Since $sub-/near$-threshold regions are not well represented by conventional transistor models and are different from those in the super-threshold region, DML modeling and sizing using a transregional model are presented. The transregional model is used to better capture DML behavior at low voltages. It also serves as a basis for the extraction of DML logical effort parameters and DML sizing methodology development under low voltage. DML robustness and silicon measurements of a benchmark conclude the chapter. DML energy–delay tradeoffs and optimization are investigated in detail in Chap. 5. In particular, the critical-path DML approaches are presented to optimize both energy efficiency and performance. Practical challenges, such as critical-path timing violations, are discussed, and DML energy–delay tradeoffs are presented using a Carry Look-Ahead adder benchmark example. Chapter 6 introduces the reader to a number of approaches for DML control. The first consists of a coarse-grain DML mode selection controller which is followed by a practical example of an energy-efficient dual-mode[2] (DM^2) system architecture. Then, an approach to fine-grain DML mode selection control is shown on an example of a CLA adder, followed by an example of a DML multiplier accumulator (MAC) with a self-adjustment mechanism, designed and fabricated in 28 nm FD-SOI. As mentioned, compatibility of a logic family to standard design flow is crucial. However, in many cases, including DML, the adaptation of a logic family to a conventional flow and standard design tools is far from straightforward. This topic is analyzed in Chaps. 7 and 8. In particular, standard design flow challenges with respect to DML are discussed, DML library characterization is presented, and a design with DML using standard flow is shown using simple benchmarks. Finally, a simple approach to DML synthesis is provided. In Chap. 9 we analyze and evaluate DML use in advanced technology nodes with advanced capabilities (namely, the FD-SOI 28 nm process). Chapter 10 outlines future challenges.

This book consolidates and combines results from a long and extensive research activity [92–104], conducted by the editors along with other faculty members, postdoctoral fellows, and PhD and master's students from Bar-Ilan (BIU) and Ben-Gurion (BGU) Universities, Israel, EPFL Switzerland, as well as the University of Calabria, Italy.

References

1. J. Rabaey, *Low Power Design Essentials* (Springer, Berlin, 2009)
2. K. Roy, S.C. Prasad, *Low-power CMOS VLSI Circuit Design* (Wiley, New York, 2009)

3. J.M. Rabaey, M. Pedram, *Low Power Design Methodologies*, vol. 336 (Springer, Berlin, 2012)
4. D. Flynn, R. Aitken, A. Gibbons, K. Shi, *Low Power Methodology Manual: For System-on-chip Design* (Springer, Berlin, 2007)
5. B.H. Calhoun, Y. Cao, X. Li, K. Mai, L.T. Pileggi, R.A. Rutenbar, K.L. Shepard, Digital circuit design challenges and opportunities in the era of nanoscale CMOS. Proc. IEEE **96**, 343–365 (2008)
6. B.H. Calhoun, J. Bolus, S. Khanna, A.D. Jurik, A.C. Weaver, T.N. Blalock, Sub-threshold operation and cross-hierarchy design for ultra low power wearable sensors, in *Proceedings of the 2009 IEEE International Symposium on Circuits and Systems* (IEEE, New York, 2009), pp. 1437–1440
7. D. Markovic, C.C. Wang, L.P. Alarcon, T.-T. Liu, J.M. Rabaey, Ultralow-power design in near-threshold region. Proc. IEEE **98**(2), 237–252 (2010)
8. D. Bol, D. Kamel, D. Flandre, J.-D. Legat, Nanometer MOSFET effects on the minimum-energy point of 45 nm subthreshold logic, in *Proceedings of the 2009 ACM/IEEE International Symposium on Low Power Electronics and Design* (ACM, New York, 2009), pp. 3–8
9. T. Jang, G. Kim, B. Kempke, B. Henry, N. Chiotellis, C. Pfeiffer, A. Grbic, D. Sylvester, D. Blaauw, Circuit and system designs of ultra-low power sensor nodes with illustration in a miniaturized GNSS logger for position tracking: Part II—Data communication, energy harvesting, power management, and digital circuits. IEEE Trans. Circuits Syst. I, Reg. Papers **64**(9), 2250–2262 (2017)
10. M. Alioto, *Enabling the Internet of Things: From Integrated Circuits to Integrated Systems* (Springer, Berlin, 2017)
11. M. Pedram, S. Nazarian, Thermal modeling, analysis, and management in VLSI circuits: Principles and methods. Proc. IEEE **94**(8), 1487–1501 (2006)
12. S. Jeong, Z. Foo, Y. Lee, J.-Y. Sim, D. Blaauw, D. Sylvester, A fully-integrated 71 nW CMOS temperature sensor for low power wireless sensor nodes. IEEE J. Solid State Circuits **49**(8), 1682–1693 (2014)
13. A. Wang, A. Chandrakasan, A 180-mV subthreshold FFT processor using a minimum energy design methodology. IEEE J. Solid State Circuits **40**(1), 310–319 (2005)
14. L.P. Alarcón, T.-T. Liu, M.D. Pierson, J.M. Rabaey, Exploring very low-energy logic: a case study. J. Low Power Electron. **3**(3), 223–233 (2007)
15. D. Blaauw, et-al., Razor II: in situ error detection and correction for PVT and SER tolerance, in *Proceedings of the 2008 IEEE International Solid-State Circuits Conference (ISSCC)* (2008), pp. 400–622
16. Y. Wu, S. Thomson, H. Sun, D. Krause, S. Yu, G. Kurio, Free razor: a novel voltage scaling low-power technique for large SoC designs. IEEE Trans. Very Large Scale Integr. VLSI Syst. **23**(11), 2431–2437 (2015)
17. G. de-Steel, F. Stas, T. Gurné, F. Durant, C. Frenkel, A. Cathelin, D. Bol, SleepTalker: a ULV 802.15. 4a IR-UWB transmitter SoC in 28-nm FDSOI achieving 14 pJ/b at 27 Mb/s with channel selection based on adaptive FBB and digitally programmable pulse shaping. IEEE J. Solid State Circuits **52**(4), 1163–1177 (2017)
18. L. Benini, G. De Micheli, State assignment for low power dissipation. IEEE J. Solid State Circuits **30**(3), 258–268 (1995)
19. Y. Xia, A.E.A. Almaini, Genetic algorithm based state assignment for power and area optimisation. IEE Proc. Comput. Digit. Tech. **149**(4), 128–133 (2002)
20. L. Xie, P. Qiu, Q. Qiu, Partitioned bus coding for energy reduction, in *Proceedings of the 2005 Asia and South Pacific Design Automation Conference* (ACM, New York, 2005), pp. 1280–1283
21. B.H. Calhoun, A.P. Chandrakasan, Ultra-dynamic voltage scaling (UDVS) using sub-threshold operation and local voltage dithering. IEEE J. Solid State Circuits **41**(1), 238–245 (2006)
22. B. Zhai, D. Blaauw, D. Sylvester, K. Flautner, The limit of dynamic voltage scaling and insomniac dynamic voltage scaling. IEEE Trans. Very Large Scale Integr. VLSI Syst. **13**(11), 1239–1252 (2005)

23. J. Shinde, S.S. Salankar, Clock gating—A power optimizing technique for VLSI circuits, in *Proceedings of the 2011 Annual IEEE India Conference (INDICON)* (IEEE, New York, 2011), pp. 1–4
24. L. Li, W. Wang, K. Choi, S. Park, M.-K. Chung, SeSCG: selective sequential clock gating for ultra-low-power multimedia mobile processor design, in *Proceedings of the 2010 IEEE International Conference on Electro/Information Technology (EIT)* (IEEE, New York, 2010), pp. 1–6
25. W. Shen, Y. Cai, X. Hong, J. Hu, An effective gated clock tree design based on activity and register aware placement. IEEE Trans. Very Large Scale Integr. VLSI Syst. **18**(12), 1639–1648 (2010)
26. H. Mahmoodi, V. Tirumalashetty, M. Cooke, K. Roy, Ultra low-power clocking scheme using energy recovery and clock gating. IEEE Trans. Very Large Scale Integr. VLSI Syst. **17**(1), 33–44 (2009)
27. R. Bhutada, Y. Manoli, Complex clock gating with integrated clock gating logic cell, in *Proceedings of the International Conference on Design and Technology of Integrated Systems in Nanoscale Era (2007 DTIS)* (IEEE, New York, 2007), pp. 164–169
28. P.K. Pal, R.S. Rathore, A.K. Rana, G. Saini, New low-power techniques: leakage feedback with Stack and Sleep stack with keeper, in *Proceedings of the 2010 International Conference on Computer and Communication Technology (ICCCT)* (IEEE, New York, 2010), pp. 296–301
29. S. Dropsho, V. Kursun, D.H. Albonesi, S. Dwarkadas, E.G. Friedman, Managing static leakage energy in microprocessor functional units, in *Proceedings of the 35th Annual IEEE/ACM International Symposium on Microarchitecture, 2002 (MICRO-35)* (IEEE, New York, 2002), pp. 321–332
30. S. Jeong, I. Lee, D. Blaauw, D. Sylvester, A 5.8 nW CMOS Wake-Up Timer for Ultra-Low-Power Wireless Applications. IEEE J. Solid State Circuits **50**(8), 1754–1763 (2015)
31. T. Jang, M. Choi, S. Jeong, S. Bang, D. Sylvester, D. Blaauw, A 4.7 nW 13.8 ppm/° C self-biased wakeup timer using a switched-resistor scheme, in *Proceedings of the 2016 IEEE International Solid-State Circuits Conference (ISSCC)* (2016), pp. 102–102
32. S. Badel, Y. Leblebici, Breaking the power-delay tradeoff: design of low-power high-speed MOS current-mode logic circuits operating with reduced supply voltage, in *Proceedings of the IEEE International Symposium on Circuits and Systems 2007 (ISCAS 2007)* (IEEE, New York, 2007), pp. 1871–1874
33. A. Inoue, V.G. Dklobdzija, W.W. Walker, M. Kai, T. Izawa, A low power SOI adder using reduced-swing charge recycling circuits, in *Proceedings of the 2001 IEEE International Solid-State Circuits Conference, 2001. Digest of Technical Papers (ISSCC)* (IEEE, New York, 2001), pp. 316–317
34. M. Li, C.-I. Ieong, M.-K. Law, P.-I. Mak, M.-I. Vai, S. -H. Pun, R.-P. Martins, Energy Optimized Subthreshold VLSI Logic Family With Unbalanced Pull-Up/Down Network and Inverse Narrow-Width Techniques. IEEE Trans. Very Large Scale Integr. VLSI Syst. **23**(12), 3119–3123 (2015)
35. H.-P. Keil, M. Momeni, A. Guntoro, A.G. Ortiz, M. Glesner, A novel leakage-estimation method for input-vector control, in *Proceedings of the IEEE Asia Pacific Conference on Circuits and Systems, 2008 (APCCAS 2008)* (IEEE, New York, 2008), pp. 570–573
36. H. Jeon, Y.-B. Kim, M. Choi, A novel technique to minimize standby leakage power in nanoscale CMOS VLSI, in *Proceeding of the IEEE Instrumentation and Measurement Technology Conference, 2009 (I2MTC'09)* (IEEE, New York, 2009), pp. 1372–1375
37. S. Mukhopadhyay, C. Neau, R.T. Cakici, A. Agarwal, C.H. Kim, K. Roy, Gate leakage reduction for scaled devices using transistor stacking. IEEE Trans. Very Large Scale Integr. VLSI Syst. **11**(4), 716–730 (2003)
38. J.-L. Kuo, H. Wang, A 24 GHz CMOS power amplifier using reversed body bias technique to improve linearity and power added efficiency, in *Proceedings of the 2012 IEEE MTT-S International Microwave Symposium Digest (MTT)* (IEEE, New York, 2012), pp. 1–3

39. L. Xiao, C. Liu, Y. Sun, A novel adaptive reverse body bias technique to minimize standby leakage power and compensate process and temperature variations, in *Cross Strait Quad-Regional Radio Science and Wireless Technology Conference (CSQRWC 2011)*, vol. 2 (IEEE, New York, 2011), pp. 1565–1568

40. K.K. Kim, Y.-B. Kim, Optimal body biasing for minimum leakage power in standby mode, in *Proceedings of the IEEE International Symposium on Circuits and Systems, 2007 (ISCAS 2007)* (IEEE, New York, 2007), pp. 1161–1164

41. K.K. Kim, Y.-B. Kim, Optimal body biasing for minimum leakage power in standby mode, in *Proceedings of the IEEE International Symposium on* Circuits and Systems, 2007 (ISCAS 2007) (IEEE, New York, 2007), pp. 1161–1164

42. W. Zhao, Y. Ha, M. Alioto, Novel self-body-biasing and statistical design for near-threshold circuits with ultra energy-efficient AES as case study. IEEE Trans. Very Large Scale Integr. VLSI Syst. **23**(8), 1390–1401 (2015)

43. H.-S. Won, K.-S. Kim, K.-O. Jeong, K.-T. Park, K.-M. Choi, J.-T. Kong, An MTCMOS design methodology and its application to mobile computing, in *Proceedings of the 2003 International Symposium on Low Power Electronics and Design, 2003 (ISLPED'03)* (IEEE, New York, 2003), pp. 110–115

44. Z. Liu, V. Kursun, Characterization of wake-up delay versus sleep mode power consumption and sleep/active mode transition energy overhead tradeoffs in MTCMOS circuits, in *Proceeding of the 51st Midwest Symposium on Circuits and Systems, 2008 (MWSCAS 2008)* (IEEE, 2008), pp. 362–365

45. J.P. Halter, F.N. Najm, A gate-level leakage power reduction method for ultra-low-power CMOS circuits, in *Proceedings of the IEEE 1997 Custom Integrated Circuits Conference, 1997* (IEEE, 1997), pp. 475–478

46. A.P. Chandrakasan, S. Sheng, R.W. Brodersen, Low-power CMOS digital design. IEICE Trans. Electron. **75**(4), 371–382 (1992)

47. G. Schrom, S. Selberherr, Ultra-low-power CMOS technologies, in *Proceeding of the International Semiconductor Conference 1996*, vol. 1 (IEEE, New York, 1996), pp. 237–246

48. A.P. Chandrakasan, R.W. Brodersen, Minimizing power consumption in digital CMOS circuits. Proc. IEEE **83**(4), 498–523 (1995)

49. S.-M.S. Kang, Elements of low power design for integrated systems, in *Proceedings of the 2003 International Symposium on Low Power Electronics and Design, 2003 (ISLPED'03)* (IEEE, New York, 2003), pp. 205–210

50. N. Verma, A.P. Chandrakasan, A 256 kb 65 nm 8t subthreshold SRAM employing sense-amplifier redundancy. IEEE J. Solid State Circuits **43**(1), 141–149, 2008

51. H.-S. Won, K.-S. Kim, K.-O. Jeong, K.-T. Park, K.-M. Choi, J.-T. Kong, Hyperdimensional computing exploiting carbon nanotube FETs, Resistive RAM, and their monolithic 3D integration. IEEE J. Solid State Circuits **53**(11), 3183–3196 (2018)

52. R. Patel, X. Guo, Q. Guo, E. Ipek, E.-G. Friedman, Reducing switching latency and energy in STT-MRAM caches with field-assisted writing. IEEE Trans. Very Large Scale Integr. VLSI Syst. **24**(1), 129–138 (2016)

53. N.H. Weste, K. Eshraghian, *Principles of CMOS VLSI design*, vol. 188 (Addison-Wesley, New York, 1985)

54. W. Al-Assadi, A.P. Jayasumana, Y.K. Malaiya, Pass-transistor logic design. Int. J. Electron. Theor. Exp. **70**(4), 739–749 (1991)

55. I.S. Abu-Khater, A. Bellaouar, M.I. Elmasry, Circuit techniques for CMOS low-power high-performance multipliers. IEEE J. Solid State Circuits **31**(10), 1535–1546 (1996)

56. R. Zimmermann, W. Fichtner, Low-power logic styles: CMOS versus pass-transistor logic. IEEE J. Solid State Circuits **32**(7), 1079–1090 (1997)

57. K. Yano, Y. Sasaki, K. Rikino, K. Seki, Top-down pass-transistor logic design. IEEE J. Solid State Circuits **31**(6), 792–803 (1996)

58. M. Anis, M. Allam, M. Elmasry, Impact of technology scaling on CMOS logic styles. IEEE Trans. Circuits Syst. II Analog Digit. Signal Process. **49**(8), 577–588 (2002)

59. S.-F. Hsiao, M.-Y. Tsai, C.-S. Wen, Transistor sizing and layout merging of basic cells in pass transistor logic cell library, in *IEEE International Symposium on VLSI Design, Automation and Test, 2008 (VLSI-DAT 2008)* (IEEE, New York, 2008), pp. 89–92

60. S. Kvatinsky, et al., Memristor-based material implication (IMPLY) logic: design principles and methodologies. IEEE Trans. Very Large Scale Integr. VLSI Syst. **22**(10), 2054–2066 (2014)

61. S. Kvatinsky, et-al., MAGIC—Memristor-aided logic. IEEE Trans. Circuits Syst. II Express Briefs **61**(11), 895–899 (2014)

62. J.-S. Friedman, A.-V. Sahakian, Complementary magnetic tunnel junction logic. IEEE Trans. Electron. Devices **61**(4), 12070–1210 (2014)

63. M.W. Allam, M.H. Anis, M.I. Elmasry, High-speed dynamic logic styles for scaled-down CMOS and MTCMOS technologies, in *Proceedings of the 2000 International Symposium on Low Power Electronics and Design* (ACM, New York, 2000), pp. 155–160

64. N.F. Goncalves, H. De Man, NORA: A racefree dynamic CMOS technique for pipelined logic structures. IEEE J. Solid State Circuits **18**(3), 261–266 (1983)

65. R. Hossain, *High Performance ASIC Design* (Cambridge University, Cambridge, 2008)

66. A. Morgenshtein, A. Fish, A. Wagner, Gate-diffusion input (GDI)-a novel power efficient method for digital circuits: a design methodology, in *Proceedings of the 14th Annual IEEE InternationalASIC/SOC Conference, 2001* (IEEE, New York, 2001), pp. 39–43

67. A. Morgenshtein, A. Fish, I.A. Wagner, Gate-diffusion input (GDI)-a technique for low power design of digital circuits: analysis and characterization, in *Proceedings of the IEEE International Symposium on Circuits and Systems, 2002 (ISCAS 2002)*, vol. 1 (IEEE, New York, 2002), pp. I–I

68. A. Morgenshtein, I. Shwartz, A. Fish, Gate diffusion input (GDI) logic in standard CMOS nanoscale process, in *Proceedings of the 2010 IEEE 26th Convention of Electrical and Electronics Engineers in Israel (IEEEI)* (IEEE, New York, 2010), pp. 000776–000780

69. V. Sze, A.P. Chandrakasan, A 0.4-v UWB baseband processor, in *Proceedings of the 2007 International Symposium on Low Power Electronics and Design* (ACM, New York, 2007), pp. 262–267

70. H. Soeleman, K. Roy, B.C. Paul, Robust subthreshold logic for ultra-low power operation. IEEE Trans. Very Large Scale Integr. VLSI Syst. **9**(1), 90–99 (2001)

71. B. Nikolic, V.G. Oklobdzija, V. Stojanovic, W. Jia, J.K.-S. Chiu, M.M.-T. Leung, Improved sense-amplifier-based flip-flop: design and measurements. IEEE J. Solid State Circuits **35**(6), 876–884 (2000)

72. A.-P. James, L.-R. Francis, D.-S. Kumar, Resistive threshold logic. IEEE Trans. Very Large Scale Integr. VLSI Syst. **2**(1), 190–195 (2014)

73. V.-T. Gaikwad, P.-R. Deshmukh, Design of CMOS ternary logic family based on single supply voltage, in *Proceedings of the 2015 International Conference on Pervasive Computing (ICPC)* (2015), pp. 1–6

74. J.M. Rabaey, A.P. Chandrakasan, B. Nikolic, *Digital Integrated Circuits*, vol. 2 (Prentice Hall, Englewood Cliffs, 2002)

75. K. Roy, S. Mukhopadhyay, H. Mahmoodi-Meimand, Leakage current mechanisms and leakage reduction techniques in deep-submicrometer CMOS circuits. Proc. IEEE **91**(2), 305–327 (2003)

76. A. Agarwal, S. Mukhopadhyay, A. Raychowdhury, K. Roy, C.H. Kim, Leakage power analysis and reduction for nanoscale circuits. IEEE Micro **26**(2), 68–80 (2006)

77. J. Kao, S. Narendra, A. Chandrakasan, Subthreshold leakage modeling and reduction techniques, in *Proceedings of the 2002 IEEE/ACM International Conference on Computer-Aided Design* (ACM, New York, 2002), pp. 141–148

78. B. Zhai, S. Hanson, D. Blaauw, D. Sylvester, Analysis and mitigation of variability in subthreshold design, in *Proceedings of the 2005 International Symposium on Low Power Electronics and Design* (ACM, New York, 2005), pp. 20–25

79. D. Bol, R. Ambroise, D. Flandre, J.-D. Legat, Analysis and minimization of practical energy in 45 nm subthreshold logic circuits, in *Proceedings of the IEEE International Conference on Computer Design, 2008 (ICCD 2008)* (IEEE, New York, 2008), pp. 294–300

80. C.-I. Kim, H. Soeleman, K. Roy, Ultra-low-power DLMS adaptive filter for hearing aid applications. IEEE Trans. Very Large Scale Integr. VLSI Syst. **11**(6), 1058–1067 (2003)
81. Y.-S. Lin, D. Sylvester, D. Blaauw, A sub-pW timer using gate leakage for ultra low-power sub-Hz monitoring systems, in *Proceedings of the IEEE Custom Integrated Circuits Conference, 2007 (CICC'07)* (IEEE, New York, 2007), pp. 397–400
82. H. Soeleman, K. Roy, B. Paul, Robust ultra-low power sub-threshold DTMOS logic, in *Proceedings of the 2000 International Symposium on Low Power Electronics and Design* (ACM, New York, 2000), pp. 25–30
83. W.M. Penney, L. Lau, *MOS Integrated Circuits: Theory, Fabrication, Design, and Systems Applications of MOS LSI* (Krieger Publishing, Florida, 1979)
84. H. Soeleman, K. Roy, B. Paul, Sub-domino logic: ultra-low power dynamic sub-threshold digital logic, in *Proceedings of the Fourteenth International Conference on VLSI Design, 2001* (IEEE, New York, 2001), pp. 211–214
85. D. Harris, M.A. Horowitz, Skew-tolerant domino circuits. IEEE J. Solid State Circuits **32**(11), 1702–1711 (1997)
86. H. Soeleman, K. Roy, Ultra-low power digital subthreshold logic circuits, in *Proceedings of the 1999 International Symposium on Low Power Electronics and Design* (ACM, New York, 1999), pp. 94–96
87. S. Thompson, I. Young, J. Greason, M. Bohr, Dual Threshold Voltages and Substrate Bias: Keys to High Performance, Low Power, 0.1 m Logic Designs, in *Proceedings of the IEEE Institute of Electrical and Electronics Symposium on VLSI Technology* (1997), pp. 69–70
88. A. Morgenshtein, V. Yuzhaninov, A. Kovshilovsky, A. Fish, Full-Swing Gate Diffusion Input logic—Case-study of low-power CLA adder design. Integration VLSI J. **47**(1), 62–70 (2014)
89. M. Alioto, G. Palumbo, Design strategies for source coupled logic gates. IEEE Trans. Circuits Systems I Fund. Theory Appl. **50**(5), 640–654 (2003)
90. A. Tajalli, E.J. Brauer, Y. Leblebici, E. Vittoz, Subthreshold source-coupled logic circuits for ultra-low-power applications. IEEE J. Solid State Circuits **43**(7), 1699–1710 (2008)
91. R.G. Dreslinski, M. Wieckowski, D. Blaauw, D. Sylvester, T. Mudge, Near-threshold computing: reclaiming Moore's law through energy efficient integrated circuits. Proc. IEEE **98**(2), 253–266 (2010)
92. A. Kaizerman, S. Fisher, A. Fish, Subthreshold dual mode logic. IEEE Trans. Very Large Scale Integr. VLSI Syst. **21**(5), 979–983 (2012)
93. I. Levi, O. Bass, A. Kaizerman, A. Belenky, A. Fish, High speed dual mode logic carry look ahead adder, in *Proceedings of the 2012 IEEE International Symposium on Circuits and Systems (ISCAS)* (IEEE, New York, 2012), pp. 3037–3040
94. I. Levi, A. Kaizerman, A. Fish, Low voltage dual mode logic: model analysis and parameter extraction. Microelectron. J. **44**(6), 553–560 (2013)
95. I. Levi, A. Fish, Dual mode logic—design for energy efficiency and high performance. IEEE Access **1**, 258–265 (2013)
96. I. Levi, A. Belenky, A. Fish, Logical effort for cmos-based dual mode logic gates. IEEE Trans. Very Large Scale Integr. VLSI Syst. **22**(5), 1042–1053 (2013)
97. A. Fish, A. Kaizerman, S. Fisher, I. Levy, Device and Method for Dual-mode Logic (2014). US Patent 8,901,965
98. R. Taco, I. Levi, M. Lanuzza, A. Fish, Evaluation of dual mode logic in 28 nm FD-SOI technology, in *Proceedings of the 2017 IEEE International Symposium on Circuits and Systems (ISCAS)* (IEEE, New York, 2017), pp. 1–4
99. V. Yuzhaninov, I. Levi, A. Fish, Design flow and characterization methodology for dual mode logic. IEEE Access **3**, 3089–3101 (2015)
100. R. Taco, I. Levi, M. Lanuzza, A. Fish, An 88-fj/40-MHZ [0.4v]–0.61-pj/1-GHZ [0.9v] dual-mode logic 8 × 8 bit multiplier accumulator with a self-adjustment mechanism in 28-nm FD-SOI. IEEE J. Solid State Circuits **54**(2), 560–568 (2018)
101. L. Moyal, I. Levi, A. Teman, A. Fish, Synthesis of dual mode logic. Integration **55**, 246–253 (2016)

102. I. Levi, A. Albeck, A. Fish, S. Wimer, A low energy and high performance DM^2 adder. IEEE Trans. Circuits Syst. I Regul. Pap. **61**(11), 3175–3183 (2014)
103. R. Taco, I. Levi, M. Lanuzza, A. Fish, Live demo: an 88FJ/40 MHZ [0.4v]–0.61 pj/1ghz [0.9v] dual mode logic 8 × 8-bit multiplier accumulator with a self-adjustment mechanism in 28 nm fd-soi, in *Proceedings of the 2019 IEEE International Symposium on Circuits and Systems (ISCAS)* (IEEE, New York, 2019), pp. 1–1
104. A. Fish, A. Kaizerman, S. Fisher, I. Levy, Device and method for dual-mode logic (2014). US Patent 8,901,965

Chapter 2
Introduction to Dual Mode Logic (DML)

This chapter discusses the concept behind DML. It presents DML basic architectures at the circuit level and describes the two modes of DML operation in detail. Specifically, it elaborates on the range of device-level topologies to construct a DML gate and the valid DML gate-level combinations. This is followed by a short discussion on the rationales, advantages, and disadvantages of each topology, as well as DML in general. In this chapter, we mainly focus on speed (performance) and energy consumption as evaluation metrics and compare DML designs to standard CMOS designs.

2.1 DML Concept and Transistor-Level Architecture

The main rationale guiding DML is to provide high-level energy–delay (E–D) optimization flexibility [1–3] at design time and runtime. The innovative capability of DML gates to switch between different operational modes at the gate level on-the-fly enables the circuit to respond to changing workloads and system states in real time. The DML paradigm makes it feasible to implement digital circuits that dissipate less energy while improving performance and reducing area at the same time. All these gains can be implemented without significant compromise in reliability.

DML operates in what are termed the *static* mode and the *dynamic* mode. In the static mode, DML gates consume very low energy, with some performance degradation compared to standard CMOS gates. On the other hand, dynamic DML gate operation exhibits very high performance at the expense of increased energy dissipation. A DML basic gate is based on a static logic family gate, e.g., a conventional CMOS gate, and an additional transistor. Although DML gates have a very simple and intuitive structure, they need an unconventional sizing scheme to achieve the desired functionality, as discussed below [1, 3].

© Springer Nature Switzerland AG 2021
I. Levi, A. Fish, *Dual Mode Logic*, https://doi.org/10.1007/978-3-030-40786-5_2

A basic DML gate architecture is made up of an un-clocked static gate, e.g., CMOS, and an additional transistor $M1$, whose gate is connected to a global clock signal [1], as shown in Fig. 2.1. This chapter discusses DML gates whose static gate implementation is based on conventional CMOS targeting a simple introduction (although there are other alternatives). A DML gate implementation can be either *Type-A* and *Type-B*, as shown in Fig. 2.1(a–d), respectively. In the static DML mode of operation (static mode), the $M1$ transistor is off by applying the high clock signal for the *Type-A* and the low $\overline{\text{Clk}}$ for the *Type-B* topology. This means that the gates of both topologies operate similarly to the static logic gate, in this case, CMOS. For dynamic operation of the gate (dynamic mode), the Clk is enabled for toggling, thus providing two separate phases: precharge and evaluation. During the precharge phase, the output is charged to V_{DD} in *Type-A* gates and discharged to GND in *Type-B* gates. During evaluation, the output is evaluated according to the

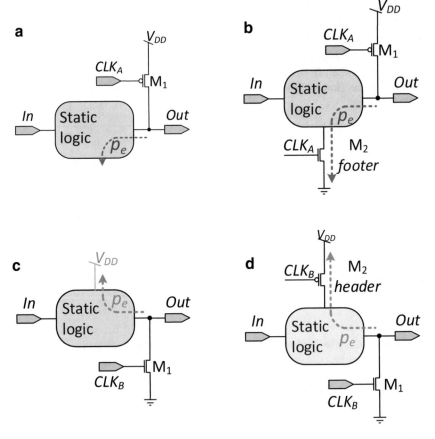

Fig. 2.1 Basic DML gate topologies: (**a**) *Type-A* footless. (**b**) *Type-A* footed. (**c**) *Type-B* footless. (**d**) *Type-B* footed. p_e traces denotes the evaluation paths for each topology

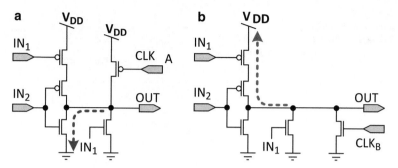

Fig. 2.2 DML NOR$_2$. (**a**) Efficient *Type-A* gate. (**b**) Less-efficient *Type-B* gate

values at the gate inputs, as is done in the NORA/np-CMOS implementations [4, 5]. Studies have confirmed that DML gates exhibit very robust operation in both the static and dynamic modes under process variations and at low supply voltages [1–3]. Dynamic mode robustness is mainly achieved through the intrinsic active restorer (pull-up network in *Type-A* and pull-down network in *Type-B*). This restorer can, however, also sustain glitches, charge leakage, and charge sharing, which are known sensitivities of standard dynamic logic families. Since DML gates have a topology which is very similar to CMOS, the design of a basic DML gate is very simple: it involves "gluing" an additional transistor for the precharge phase and, in the case of a footed gate, adding an additional nMOS transistor as a footer in *Type-A* gates and a pMOS transistor as a header in *Type-B* gates.

As discussed in the previous paragraph, the most efficient DML gates are typically the ones with a precharge (or predischarge) transistor connected in parallel to a group of serially stacked transistors that are minimally sized (either pull-up or pull-down). Therefore, the evaluation network is usually dominated by parallel paths which contribute to a very fast evaluation period (small evaluation path resistance and reduced output capacitance). In general, the designer is not obligated to use these guidelines and the precharge transistor can be placed in parallel to a parallel path network, but this will result in relatively slow DML gates (compared to the opposite type). Hence, to fully exploit the DML advantages, specific gates are better utilized in certain types. Figure 2.2 illustrates this principle, where a DML *Type-A* NOR$_2$ gate is very fast in comparison to a DML *Type-B* NOR$_2$ gate.

The unique sizing of the DML gate transistors is the key factor in achieving low energy consumption in the static DML mode (where the topology of the gate is identical to the static gate). This sizing is also responsible for the reduction of all the capacitances of the gate. Similarly, the unique transistor sizing and utilization of an appropriate topology enables evaluation by a low resistance network, thus achieving fast operation in the dynamic mode.

An intuitive visualization of the tradeoff inherently related to DML is shown in Fig. 2.3. Energy efficiency in the static DML mode comes at the cost of slower operation (low energy and low performance, left side of the fulcrum). By contrast,

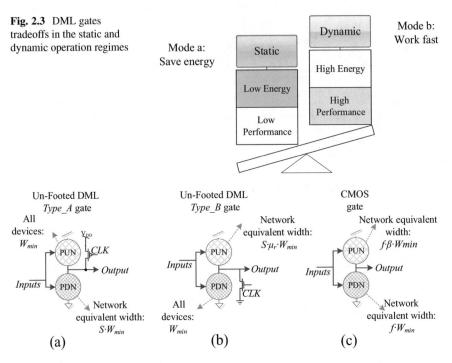

Fig. 2.3 DML gates tradeoffs in the static and dynamic operation regimes

Fig. 2.4 DML characteristics: (**a**) general *un-footed Type-A* CMOS-based DML gate with detailed sizing, (**b**) general *un-footed Type-B* CMOS-based DML gate with detailed sizing, and (**c**) general CMOS gate with detailed sizing

the dynamic mode is characterized by high performance, although with increased energy consumption (high energy and high performance, right side of the fulcrum). These tradeoffs enable very high levels of flexibility at the system level, as described in detail in the following chapters.

Figure 2.4a and b shows the sizing of CMOS-based DML gates in *Type-A* and *Type-B*, respectively. These (exemplary) DML gates are optimized for dynamic operation for which the equivalent widths of the evaluation networks are $S\dot{W}_{min}$ and $S\mu_r\dot{W}_{min}$ for the *Type-A* and *Type-B*, respectively. Figure 2.4c depicts the conventional sizing of a standard CMOS gate where W_{MIN} is the minimal transistor width, β is the PUN to PDN inherent upsizing factor (which mainly comes from hole and electron mobility ratio), and f is the general upsizing factor of the gate [3, 6, 7]. Note that although the upsizing factors S and f of DML and CMOS gates have the same meaning, they are calculated in a slightly different way, as described in the next chapter. The in/out capacitances of the DML gates are significantly smaller than CMOS gates, as a result of the utilization of minimal width transistors in the complementary networks. The size of the precharge transistor remains the same $S \times W_{MIN}$ to preserve a fast precharge period, despite the output load upsized gate, where S again is the evaluation network upsizing factor. More details are provided in [3].

As is the case for other dynamic families, DML gates can be designed with or without a footer for *Type-A* DML (or a header for *Type-B* DML). Figure 2.1b and d illustrates the footed *Type-A* and the headed *Type-B* DML gates, respectively. These topologies are explained in detail in [8]. They enable the successful precharge of a cascaded topology of standard static gates/sequential devices to DML logic. The footer is also used to cut down on the precharge time by eliminating the ripple effect of the data advancing through the cascaded gates and by allowing for faster precharge. Many features of DML gate sizing, as well as the preferred set of gates for *Type-A* and *Type-B* topologies, have been analyzed and discussed. Optimization to find the best upsizing parameters for a network (pull-up and pull-down), in the sense of fast load driving, has been explored using the logical effort (LE) method [3]. The main advantage to DML is that while presenting very high performance in the dynamic mode through its sizing, the same topology also leads to considerable energy efficiency in the static mode as compared to a conventional CMOS.

2.2 DML Advantages

2.2.1 Robust Operation, Inherent Keeper, and High Performance

In addition to its unique ability to switch between its two operating modes, DML nodes that run in the dynamic mode have a number of major advantages over conventional dynamic nodes which are due to the sizing and topology of DML transistors, as discussed in this section.

Throughout this book, we focus on DML gate optimization to improve speed in the dynamic mode, while still maintaining a very low minimum energy point during static operation. We consider this form of optimization to be the most interesting/practical approach. Thus, although we primarily optimize for performance, DML systems still provide an overall minimum energy operation, as compared to their CMOS counterparts. Unlike CMOS gates, each DML gate can be implemented in two ways, only one of which is energy efficient. Since performance in the dynamic mode is mainly determined by evaluation speed, the preferred topology is where the precharge transistor is placed in parallel to the stacked transistors, i.e., NOR in *Type-A* is preferred over NAND, and NAND in *Type-B* is preferred over NOR. In this case, the evaluation is performed through the parallel transistors and therefore is faster. As noted above, minimum-sized transistors are utilized in the active self-restore network. This allows capacitance reduction at the gate output, especially for gates with a large fan-in. The strength of the evaluation network is set to be equivalent to one minimum-sized nMOS transistor, similar to the standard CMOS methodology. It is worth noting that all gates can be designed as either *Type-A* or *Type-B*, by ignoring the optimization guidelines mentioned above (e.g., NOR gates in *Type-A* and NAND gates in *Type-B*). The optimal design methodology when designing with DML gates is thus to connect *Type-A* and *Type-B* gates, exactly as in np-CMOS gates. Even though this design methodology will allow maximum performance by minimizing area and power efficiency, it is also possible to connect gates of the same type using an inverter, and buffering between them, as in domino logic. Connecting gates of the same type without inverters is also possible when a footer/header is used at each stage; however, this structure will cause glitching after the end of precharge until the evaluation data ripples through the chain. These are standard challenges when designing with dynamic gates [9], but unlike standard dynamic logic, DML's inherent keeper helps recover the logical values.

DML's key advantages lie in its abilities to scale with the technology and with the power supply voltage (these features are covered throughout this book). Due to the active inherent keeper (pull-up network in a *Type-A* gate and pull-down network in a *Type-B* gate), DML can operate dynamically down to subthreshold voltage while still providing the same (and sometimes even more) benefits as in super-threshold operation. For the same reasons, it solves charge-sharing and leakage issues through its clocked devices, unlike other dynamic families which proved to be not robust to scaling and lowering of the supply voltage.

These features constitute even stronger reasons to utilize DML for low-voltage operation with high(er) performance. Ultra-low/subthreshold operations are still not widely adapted, because of their significant degradation in performance. Domino low-voltage logic was suggested as a possible solution [10], but its high sensitivity to process variations is a real drawback. Even dynamic logic with process scaling is being abandoned in the super-threshold regime, given its very low yield and logic failures. However, the main issues and shortcomings of low-voltage dynamic logic are elegantly solved or simply avoided when using DML. Charge leakage and charge sharing are no longer problematic in DML, since the complementary part acts as a keeper and restores the logical level, without requiring a high-power-consuming bleeder or an area and power-consuming keeper. The ability to properly restore the logical levels also avoids the back gate coupling issue. In the following chapters, we discuss the optimization of the DML gates and derive the transistor sizes for mid-low-voltage operations.

2.3   DML: The Best of Both Worlds

The performance of most digital circuits and systems is determined by the delay of the Critical Path (CP). Even though standard synthesis tools are geared to design logic blocks without CP [11–13] (i.e., equalized path delay), the slack from the targeted clock frequency still exists and needs to be remedied during the design phase. Many methods have been proposed to address these slacks. These include adaptive voltage scaling with a CP emulator circuit [14], multi-oxide thickness-driven threshold voltages, multichannel lengths for energy reduction in the non-CPs, and performance boost in the CPs [15, 16]. Meijer et al. and Liu et al. applied a body bias on a non-CP to improve energy consumption and increase performance of the

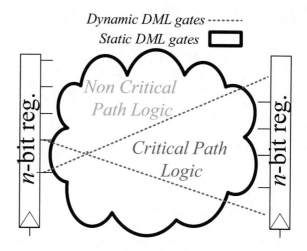

Fig. 2.5 A circuit pre-processed and mapped for CPs and non-CPs where the former operates in the dynamic DML mode and the latter in the static energy-efficient DML mode

CPs, respectively [8, 17], and many other more recent techniques are discussed in Sect. 1.1. Although these methods solve the critical path slack problem, in most cases they also lead to a significant increase in energy consumption or resource overhead.

The overall approach to DML design aims to meet the delay requirements of CPs while also lowering the overall energy consumption of the design by utilizing the powerful modularity of DML. First, the design is analyzed to locate the CPs; then during runtime the on-the-fly modularity of DML is utilized to operate these paths in the boosted (dynamic) performance mode. The noncritical paths are operated in the low-energy static DML mode which does not affect the performance of the design. Since in most design cases the majority of the gates are not on the CPs, the increase in energy consumption of the critical paths is negligible compared to the general circuit consumption. DML static gates dissipate less power than their CMOS counterparts, resulting in less power dissipation of the whole design. These features are illustrated in Fig. 2.5.

In terms of the big picture of what we aim to achieve with DML, the utilization of DML makes it possible to extend *classic* E–D curves and operations in both the energy and performance directions, i.e., the goal is to lower the minimum energy point (MEP) of the design while still providing a lower minimum delay point (MDP) or a more *worthwhile* tradeoff between the two. In other words, with the same design, by switching from the static mode to the dynamic mode, we can extend both the MEP and the MDP as compared to conventional static CMOS design. This idea is schematically illustrated in Fig. 2.6. The figure presents the traditional E–D space, with its design characteristics (energy per operation and worst-case delay). Each point on the graphs represents operations under a different power supply voltage. For standard CMOS designs with high-supply voltage, the MDP is achieved by trading off energy characteristics. On the other hand, when lowering the power

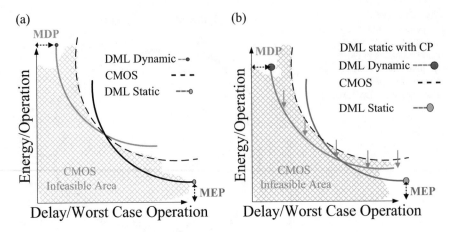

Fig. 2.6 E–D space for CMOS, the DML dynamic mode, and the DML static mode as a function of V_{DD}. (**a**) Entire design either in Dynamic or Static DML mode (**b**) Only the critical-path is operated in the DML Dynamic mode while the rest of the design is operated in the Static mode

supply voltage, performance is reduced as well as the energy reaching the MEP at a given voltage. In DML, when the entire design is operated in the dynamic mode (orange curve), the MDP will improve (although with increased energy), as shown in Fig. 2.6a. When the entire design is operated in the static mode (dark blue curve), the MEP will improve (for more details on the power supply voltage dependency, see the next two chapters). As shown in the illustration of Fig. 2.6b, the main goal/challenge of DML is to find a "combined" mode at the system level that achieves the best of both worlds. The natural strategy is to only operate critical elements/paths in the dynamic mode, while the rest (most) of the system is operated in the static mode. This is highlighted by the light blue curve in the figure. Note that depending on the control strategy at the architectural level, the system will alternate between the dark blue curve (when high performance is not needed) and the light blue curve (when it is). As in the optimization process (in most cases), we give more priority to performance. The MDP is the best possible for DML, whereas the MEP is typically suboptimal due to the energy overheads of the control circuitry.

Now that the reader is slightly better acquainted with DML, we take this opportunity to restate the main objective of the following chapters. We detail the basic properties of the DML family, the optimization criteria, and evaluation metrics (e.g., energy consumption, performance, robustness). We provide several examples of the systematic utilization of DML at the system level. Finally, we show how to control DML modes of operation and present both data-driven and external signal-driven control schemes.

References

1. A. Kaizerman, S. Fisher, and A. Fish, Subthreshold dual mode logic. IEEE Trans. Very Large Scale Integr. (VLSI) Syst. **21**(5), 979–983 (2013)
2. I. Levi, A. Kaizerman, A. Fish, Low voltage dual mode logic: Model analysis and parameter extraction. Microelectron. J. **44**(6), 553–560 (2013)
3. I. Levi, A. Belenky, A. Fish, Logical effort for cmos-based dual mode logic gates. IEEE Trans. Very Large Scale Integr. (VLSI) Syst. **22**(5), 1042–1053 (2014)
4. J.M. Rabaey, A.P. Chandrakasan, B. Nikolic, *Digital Integrated Circuits*, vol. 2 (Prentice Hall, Englewood Cliffs, 2002)
5. N.F. Goncalves, H. De Man, Nora: A racefree dynamic cmos technique for pipelined logic structures. IEEE J. Solid-State Circuits **18**(3), 261–266 (1983)
6. M.K. Stojčev, Jan m. rabaey, anantha chandrakasan, and borivoje nikolić: Digital integrated circuits: A design perspective, 2/e. Facta Univ. Ser. Electron. Energetics **16**(1), 155–157 (2003)
7. I.E. Sutherland, R.F. Sproull, D.F. Harris, *Logical Effort: Designing Fast CMOS Circuits* (Morgan Kaufmann, San Francisco, 1999)
8. M. Meijer, J.P. de Gyvez, Body-bias-driven design strategy for area-and performance-efficient cmos circuits. IEEE Trans. Very Large Scale Integr. (VLSI) Syst. **20**(1), 42–51 (2012)
9. R. Hossain, *High Performance ASIC Design: Using Synthesizable Domino Logic in an ASIC Flow* (Cambridge University Press, Cambridge, 2008)
10. H. Soeleman, K. Roy, B. Paul, Sub-domino logic: ultra-low power dynamic sub-threshold digital logic, in *Fourteenth International Conference on VLSI Design, 2001* (IEEE, Piscataway, 2001), pp. 211–214
11. Y. Kukimoto, M. Berkelaar, K. Sakallah, Static timing analysis, in *Logic Synthesis and Verification* (Springer, Boston, 2002), pp. 373–401
12. T. Sasao, *Switching Theory for Logic Synthesis* (Springer Science & Business Media, New York, 2012)
13. J.J. Zasio, K.C. Choy, D.R. Parham, Static timing analysis of semiconductor digital circuits. May 8 1990, US Patent 4,924,430
14. M. Elgebaly, M. Sachdev, Efficient adaptive voltage scaling system through on-chip critical path emulation, in *Proceedings of the 2004 International Symposium on Low Power Electronics and Design* (ACM, New York, 2004), pp. 375–380
15. H.I. Chen, E.K. Loo, J.B. Kuo, M.J. Syrzycki, Triple-threshold static power minimization technique in high-level synthesis for designing high-speed low-power soc applications using 90nm mtcmos technology, in *2007 Canadian Conference on Electrical and Computer Engineering* (IEEE, Piscataway, 2007), pp. 1671–1674
16. N. Sirisantana, L. Wei, K. Roy, High-performance low-power cmos circuits using multiple channel length and multiple oxide thickness, in *2000 International Conference on Computer Design, 2000. Proceedings* (IEEE, Piscataway, 2000), pp. 227–232.
17. X. Liu, S. Mourad, Performance of submicron cmos devices and gates with substrate biasing, in *The 2000 IEEE International Symposium on Circuits and Systems, 2000. Proceedings. ISCAS 2000 Geneva*, vol. 4 (IEEE, Piscataway, 2000), pp. 9–12

Chapter 3
Optimization of DML Gates

This chapter presents several techniques to achieve high-performance and/or low-energy operation of DML circuits. We introduce several optimization methodologies for DML circuits while focusing on gate-level techniques. This goal is primarily achieved by utilizing the logical effort model (LE) which was uniquely adapted to DML. We discuss several approaches which trade off the accuracy of the solution with simplicity and complexity (i.e., the complete and approximate LE models). The method is then generalized to complex gates and branches. Finally, we compare and evaluate the methods discussed.

3.1 Introduction

Logic optimization and timing estimations are basic tasks for digital circuit designers. The logical effort (LE) method was first presented by Sutherland et al.[1–4] for easy and fast evaluation and optimization of delays in CMOS logic paths. Its elegance has made the LE method a very popular tool for designing and education purposes and is the basis for several EDA tools [3–7]. Although LE is mainly used for standard CMOS logic, it has also been shown to be useful for other logic families such as the pass transistor logic (PTL) [8]. In this chapter we tackle the natural questions of the feasibility of adapting LE to optimize one or both dual mode logic (DML) modes. DML gates have a very simple, intuitive structure, but they require an unconventional sizing methodology to achieve target performance. Conventional LE methodology cannot be used with the DML family since it does not consider its unconventional sizing rules and topology. For this reason, and to keep the discussion general, in this book we refer to the LE device-sizing technique for the *CMOS-based* DML family.

The objective of this chapter is to develop a simple method for minimizing delays and achieving an optimized number of stages in logical paths containing CMOS-based DML gates. A unified LE method is introduced for the delay evaluation and

© Springer Nature Switzerland AG 2021
I. Levi, A. Fish, *Dual Mode Logic*, https://doi.org/10.1007/978-3-030-40786-5_3

optimization of logic paths constructed with DML logic gates. DML-LE solves complete (non-approximate) design problems, which can be solved numerically, and simplifies these problems to straightforward, easy to perform computational problems (with approximate and semi-approximate solutions) by applying a unified analytic model. This model estimates the minimum to maximum error under delay approximation and the error in the optimal number of stages for a given logic function. We compare DML-LE theoretical results to simulation results using the Cadence Virtuoso optimizer tool implementing standard 40 nm technology. To begin, we first lay out the development of a DML-LE model for simple inverter chains with three different levels of approximation. These methods are then compared in terms of their simplicity and accuracy. The dependence of the optimal number of stages is also described, along with an intuitive graphic visualization of the problem. DML-LE is then expanded to complex nets containing branching, and finally, the efficiency of the DML-LE theoretical optimization is examined for a standard technology process. We end with a discussion on the use and applicability of DML-LE.

3.2 Overview: Standard Logical Effort (LE) Model for a Simple CMOS Inverter Chain

LE is a simplified transistor sizing optimization method to achieve improved metrics in combinational logic. The detailed description of the conventional LE semantics can be found in many works, as indicated in the previous subchapter. These techniques were extended here to include advanced features of latest technologies such as temperature/voltage, low voltage, interconnect inclusion, energy–delay optimization, and complex cell fitting. In LE the gate normalized delay of stage i (D_i) in a gate chain can be expressed as the sum of the stage (f_i) and parasitic (p_i) efforts:

$$D_i = f_i + p_i \tag{3.1}$$

where $f_i = g_i \cdot h_i \cdot b_i$, g_i is the logical effort of the stage and h_i is the electrical effort of the stage:

$$h_i \triangleq \frac{C_{out\_i}}{C_{in\_i}} \tag{3.2}$$

where C_{out_i} and C_{in_i} are the output and input capacitance of the stage i element, respectively, and b_i is the branching effort of the stage:

$$b_i = \frac{C_{on\_path\_i} + C_{off\_path\_i}}{C_{on\_path\_i}} \tag{3.3}$$

where $C_{on\_path\_i}$ and $C_{off\_path\_i}$ denote the output capacitance portion which is associated with the logical path under optimization and the rest of the stage load (off-paths).

We use the terminology presented in [8]. The logical effort of the stage is denoted by LE_i and the electrical effort is expressed by f_i. The equation is normalized, for all parameters, to a simple inverter; therefore:

$$p_i \triangleq \frac{R_{gate\_i}}{R_{inv}} \cdot \frac{C_{D,gate\_i}}{C_{D,inv}} \tag{3.4}$$

where R_{gate_i} and $C_{D,gate\_i}$ are the gate's resistance and inherent output (drain) capacitance, respectively. R_{inv} and $C_{D,inv}$ are normalization factors. Specifically, the resistance and output capacitance of a minimum sized inverter (their exact formulation follows) are:

$$LE_i = g_i \triangleq \frac{R_{gate\_i} C_{G,gate\_i}}{R_{inv} C_{G,inv}} \tag{3.5}$$

where $C_{G,gate\_i}$ and $C_{G,inv}$ are the gate's inherent input (gate) capacitance and the input capacitance of a minimum sized inverter, respectively.

Using this terminology, the delay of a gate in stage i is:

$$D_i = t_{pd\_i} = t_{P0}(p_i * \gamma + LE_i * \beta_i * \frac{C_{on\_path\_i}}{C_{in\_gate\_i}}) = t_{P0}(p_i * \gamma + EF_i) \tag{3.6}$$

$$EF_i = LE_i \cdot b_i \cdot h_i \tag{3.7}$$

and

$$t_{P0} = \frac{0.69 R_{inv} C_{d\_inv}}{\gamma} = \frac{t_{p.inv}}{\gamma} \tag{3.8}$$

where γ is a process parameter, deduced from:

$$\frac{C_{d\_inv}}{\gamma} = C_{g\_inv} \tag{3.9}$$

In CMOS logic, the pMOS pull-up network (PUN) to nMOS pull-down network (PDN) sizing ratio is denoted by β. β parameter is due to holes and electron mobility differences. In the case of average gate resistance, the PUN and PDN resistance ratio is:

$$\beta_{opt} \approx \sqrt{\frac{R_{eqp}}{R_{eqn}}} = \sqrt{\frac{\mu_e}{\mu_p}} \tag{3.10}$$

$$R_{inv} \simeq \frac{\left(R_{eqn} + \frac{R_{eqp}}{\beta}\right)}{2} \; ; \; R_{eqp} \simeq \frac{\mu_e}{\mu_p} R_{eqn} \tag{3.11}$$

Conventional LE provides a well-explored solution for the upsizing of a given CMOS gate chain. The upsizing factors and number of gates required in the chain are constrained by the load capacitance, C_{LOAD}, logical functions, area, delay, and power requirements. First, the chain delay is estimated:

$$D = t_{pd} = \sum_{1}^{N} D_i = t_{P0} \sum_{1}^{N} (p_i * \gamma + EF_i) \tag{3.12}$$

The optimal chain sizing considers upsizing each stage by an optimal electrical effort (EF_{opt}), which is given by:

$$EF_{opt} = \sqrt[N]{PE} = \sqrt[N]{F * \prod LE_i * \prod b_i} \tag{3.13}$$

where PE is the path effort and F is the C_{LOAD} to input capacitance ratio. For a given chain, containing N CMOS gates, N is not necessarily equal to the optimal number of stages, N_{opt}. If $N < N_{opt}$, a number of inverters can be added to better fit the stage effort of the path and therefore improve its delay. For $N < N_{opt}$, EF_{opt} is given by:

$$EF_{opt} = 3.6 \text{ (for } \gamma = 1) \tag{3.14}$$

For this case, N_{opt} is given by $\sqrt[N_{opt}]{PE} = EF_{opt}$. For $N > N_{opt}$, EF_{opt} can be approximated as:

$$\sqrt[N]{PE} = EF_{opt} \tag{3.15}$$

Note that EF_{opt} (γ, p_{inv}-dependent) may not be feasible in a path where N, PE are constrained.

3.3 Logical Effort (LE) Model for a Simple DML Inverter Chain

To optimize the performance of the DML gates, we employ, modify, and approximate the LE technique [1, 2]. Although LE is a well-known method and is widely used by designers, we need a few different metrics and terminologies. The terminology used in this chapter appears in the previous subsection. The logical effort formulation of DML is very different from the conventional CMOS-LE (and domino logic LE)[1, 2] which was discussed above. This is because of the unique structure and unconventional sizing methodology of DML gates. Achieving the optimal non-approximate solution is an exhausting task. However, with minor simplifications it can be solved similarly to the standard CMOS-LE method. In this subsection, we first present a complete non-approximated LE method for DML CMOS-based gates. Although this solution is very precise, it is highly complex and distinctly not designer friendly. We thus provide two approximated solutions whose complexity is much lower, but still achieve very high precision. Finally, we discuss these approaches as they apply to DML-LE for all CMOS-based gates.[1]

3.3.1   Basic Assumptions

DML gates are designed to optimize their dynamic mode delay so that only one transition out of $Tplh$ and $Tphl$, which is part of the evaluation phase, is considered. This means that only an equivalent resistance of the pull-down network (nMOSs) plays a role in the delay optimization of $Type$-A gates, and the pull-up network (pMOSs) will be applicable to the optimization of $Type$-B gates. When designing conventional CMOS gates, the pull-up network is typically upsized with β, independently of the sizing factor EF_{opt}, which is the sizing contribution of the load driving effort. This β is the outcome of the optimal delay of an unloaded gate. Generally, β, derived for an optimal gate delay, differs from β_{sym} that achieves symmetric gate operation ($Tphl = Tplh$). However, in most technologies β is approximately equal to β_{sym} ($\beta = \beta_{sym}$) [8]. In DML, each standalone gate is not sized with β since the delay in the dynamic mode is determined by a single transition through PDN or PUN so that there is no need of symmetric transitions. Only one sizing factor, S_i, for each i stage gate impacts the evaluation network and the precharge transistor, as shown in Fig. 3.1. In the CMOS-LE method, normalization is performed on a standard CMOS inverter. DML gates are normalized to a standard minimal inverter (DML_INV) of $Type$-A, which represents the minimal standalone gate delay unit. A minimal inverter of "Type B"Ì presents an increased delay because it evaluates the data through pMOS. In this chapter we assume that each DML chain starts with $Type$-A gates, followed by $Type$-B gates (as in a NORA np-CMOS flavor [8, 9]). As mentioned in the previous subsection, γ is the fabrication technology-dependent factor that defines the transistor gate capacitance to transistor drain capacitance ratio. Typically, in most nanometer-scale processes γ is close to 1. For CMOS inverters it also describes the gate-to-drain capacitance of a single MOS transistor. However, for an all-minimal transistor width DML_INV $Type$-A or $Type$-B:

---

[1]Note that DML's dual-mode methodology can be applied *over* other static-logic families, e.g., PTL and not only CMOS; however, in this case the DML-LE approach in this chapter would not apply directly.

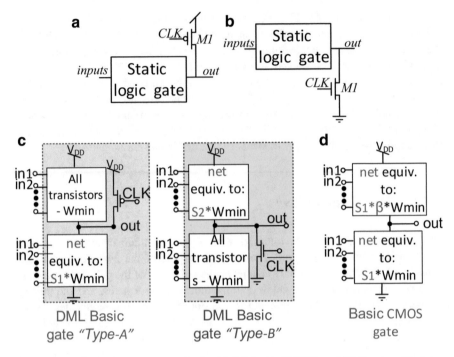

Fig. 3.1 (**a**) *Type-A* DML topology. (**b**) *Type-B* DML. (**c**) CMOS-based DML gate with sizing factors. (**d**) Standard CMOS gate with sizing factors

$$\frac{C_{d\_inv\_DML}}{C_{g\_inv\_DML}} = \frac{3C_{d\_MOS}}{2C_{g\_MOS}} \tag{3.16}$$

where C_{d_MOS} and C_{g_MOS} are a minimum-sized MOS transistor drain and gate capacitance, respectively. All of which yields:

$$\gamma' = 3\gamma/2 \tag{3.17}$$

where γ' represents the equivalent γ factor corresponding to the DML inverters.

3.3.2 Defining the Optimization Target for a Simple Inverter Chain

In order to extract the optimal sizing factors for a simple DML inverter chain, we assume a chain, shown in Fig. 3.2. The delay of a general gate *i* in the chain is given next.

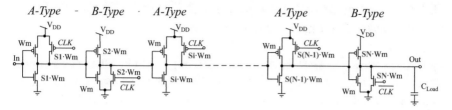

Fig. 3.2 DML inverter chain with sizing factors (W_m stands for W_{min})

$$
t_{pd\_i} = \underbrace{\frac{\ln(2) \cdot R_{min\_A} C_{D,min}}{\gamma'}}_{t_{p0\_DML}} \left(\overbrace{\frac{R_{gate}}{R_{inv}} \cdot \frac{C_{D,gate}}{C_{D,inv}}}^{p\_DML} \gamma' + \overbrace{\frac{R_{gate}}{R_{inv}} \cdot \frac{C_{G,gate}}{C_{G,inv}}}^{LE\_DML} \underbrace{\frac{C_{Load}}{C_{G,gate}}}_{f\_DML} \right),
$$
$$(3.18)$$

where R_{min_A}, $C_{D,min}$ are the resistance and drain (output) capacitance of a minimal *Type-A* DML inverter. The normalized delay of each odd gate (*Type-A*) and each even gate (*Type-B*) can be described in terms of the delay of a *Type-A* minimal DML inverter t_{p0_DML}:

$$
\begin{aligned}
t_{pd\_i\_odd} &= t_{p0\_DML} \left(\tfrac{(2s_i+1)}{3s_i} \gamma' + \tfrac{(s_{i+1}+1)}{2s_i} \right) \\
t_{pd\_i\_even} &= t_{p0\_DML} \left(\mu_{n/p} \left[\tfrac{(2s_i+1)}{3s_i} \gamma' + \tfrac{(s_{i+1}+1)}{2s_i} \right] \right)
\end{aligned}
$$
$$(3.19)$$

where $\mu_{n/p}$ is defined as μ_n/μ_p and S_i is the ith stage sizing factor (shown in Fig. 3.2). Then, assuming an even number of inverters N in the chain, the delay of the chain can be expressed by adding the delays of all the chain components:

$$
D = \sum_i t_{pd\_i} = t_{p0\_DML} \left(\begin{array}{c} \displaystyle\sum_{\substack{odd\_i \\ Type\_A}} \left(\tfrac{(2s_i+1)}{3s_i} \gamma' + \tfrac{(s_{i+1}+1)}{2s_i} \right) + \\[2ex] + \displaystyle\sum_{\substack{even\_i \\ Type\_A}} \left(\mu_{n/p} \left[\tfrac{(2s_i+1)}{3s_i} \gamma' + \tfrac{(s_{i+1}+1)}{2s_i} \right] \right) \end{array} \right).
$$
$$(3.20)$$

In the following subsections, three different solutions to the delay optimization problem are developed. The first is a complete un-approximated solution, the second is a complete approximated solution, and the third is a partially/semi-approximated solution. These solutions illustrate tradeoffs between complexity and accuracy.

3.3.3   The Complete Un-approximated Method (CS) for DML Sizing Factors of an Inverter Chain

In order to solve this problem, we differentiate Eq. (3.20) by all S_i factors of the chain and equate to 0, i.e., $\frac{dD}{ds_i} = 0$. After simplification and substituting γ', the following expression can be written for all odd $i's$ Eq. (3.21) and all even $i's$ Eq. (3.22):

$$\text{odd } i : \quad \frac{S_i}{S_{i-1}} = \frac{(\gamma + 1 + S_{i+1})}{S_i} \frac{1}{\mu_{n/p}}, \qquad (3.21)$$

$$\text{even } i : \quad \frac{S_i}{S_{i-1}} = \frac{(\gamma + 1 + S_{i+1})}{S_i} \mu_{n/p}. \qquad (3.22)$$

Typically, the first gate in the chain is composed of all minimally sized transistors and therefore $S_1 = 1$. Assuming $B = \mu_{n/p}$, $B_2 = (\gamma + 1)\mu_{n/p}$, Eq. (3.21) and (3.22) can be represented by the set of expressions as shown in Eq. (3.23). This is a set of N equations with N unknown variables; each equation is nonlinear, containing mixed variable multiplication. In general, it can be solved numerically. This type of solution is the best, most accurate solution for DML inverter chain sizing (denoted by CS for "complex solution"). However, solving it is a Herculean task, since it is much more complex than a simple CMOS-LE optimal solution which is derived with no assumptions or approximations, as was shown in Sect. 3.2. The DML CS method complexity is the outcome of the nonstandard sizing of transistors connected in parallel to the Clk-ed transistor.

$$
\begin{aligned}
S_1 &= 1 \\
0 &= B_2 S_1 - S_2{}^2 + B S_1 S_3 \\
0 &= B_2 S_2 - B^2 S_3{}^2 + B S_2 S_4 \\
0 &= B_2 S_3 - S_4{}^2 + B S_3 S_5 \\
0 &= B_2 S_4 - B^2 S_5{}^2 + B S_4 S_6 \\
&\;\vdots \;\; \vdots \;\; \vdots \;\; \vdots \;\; \vdots \\
S_N{}^2 &= B_2 S_{N-1} + B S_{N-1} S_{N+1}
\end{aligned}
\qquad (3.23)
$$

Note that the following assumptions will be used in the rest of this chapter. First, as assumed in the last subsection, the first gate of any chain is minimum-sized, i.e., $S_1 = 1$. Bear in mind that S_1 can be generalized to any possible size as a function of any input capacitance. Second, an even number of stages N is assumed. This is the result of the topology of DML chains which basically consist of *Type-B* gates following *Type-A* gates. However, the solution for a chain that has an odd number of stages can easily be derived using the same methodology.

3.3.4 The Complete Approximated Method (CA) for DML Sizing Factors of an Inverter Chain

To reduce the complexity of the LE method, a complete approximated solution which trades off complexity and accuracy is presented.

As explained Eq. (3.20) provides the general delay expression for the whole chain, assuming an even number of inverters N. The CA method assumes that the contribution of minimal transistors to the drain and gate capacitances is negligible as compared to $2S_i$ and to S_{i+1} for all stages of the chain. As shown in the following subsections, neglecting these transistors for complex gates enhances the accuracy with respect to inverters. Next, Eq. (3.20) can be rewritten by:

$$D = \sum_N D_i = t_{p0\_DML} \left(\begin{array}{l} \displaystyle\sum_{\substack{odd\_i \\ Type\_A}} \left(\frac{(2s_i\overline{+1})}{3s_i} \gamma' + \frac{(s_{i+1}\overline{+1})}{2s_i} \right) + \\[2em] \displaystyle + \sum_{\substack{even\_i \\ Type\_B}} \left(\mu_{n/p} \left[\frac{(2s_i\overline{+1})}{3s_i} \gamma' + \frac{(s_{i+1}\overline{+1})}{2s_i} \right] \right) \end{array} \right). \qquad (3.24)$$

These assumptions are only justified when the output load capacitance of the chain is large since a large load capacitance affects the sizing factors S_i. Since S_i increases as i increases along the chain, this approximation will increase in accuracy for large $i's$. After simplification, Eq. (3.24) can be rewritten as:

$$D = \sum_N D_i = t_{p0\_DML} \left(\sum_{\substack{odd\_i \\ Type\_A}} \left(\frac{2}{3} \gamma' + \frac{s_{i+1}}{2s_i} \right) + \sum_{\substack{even\_i \\ Type\_B}} \left(\mu_{n/p} \left[\frac{2}{3} \gamma' + \frac{s_{i+1}}{2s_i} \right] \right) \right).$$
$$(3.25)$$

By differentiating $dD/ds_i = 0$ and following the same procedure for all odd $i's$ and all even $i's$:

$$\text{odd } i : \quad \frac{S_i}{S_{i-1}} = \frac{S_{i+1}}{S_i} \frac{1}{\mu_{n/p}}, \qquad (3.26)$$

$$\text{even } i : \quad \frac{S_i}{S_{i-1}} = \frac{S_{i+1}}{S_i} \mu_{n/p}. \qquad (3.27)$$

The sizing factor solution to this complete approximated approach is very similar to the standard CMOS solution. Like CMOS, the upsizing factor is constant; however, all even stages are factored by an additional $\sqrt{\mu_{n/p}}$. For the N-size chain, the sizing factors can be written in series as in Table 3.1:

Table 3.1 Inverter chain sizing factors, S_i, for the CA method

S_1	S_2	S_3	S_4	S_5	S_{N-1}	S_N
1	$\sqrt{\mu_{n/p}}A^{0.5}$	A	$\sqrt{\mu_{n/p}}A^{1.5}$	A^2	$\sqrt{\mu_{n/p}}A^{(\frac{N}{2}-0.5)}$	$A^{\frac{N}{2}}$

where A is expressed in Eq. (3.29). Thus, while in CMOS the sizing factors were derived from the load to input capacitance ratio, in DML they are determined by the ratio of the first to the last sizing factors, i.e.:

$$\text{In CMOS: } F = \frac{C_{Load}}{C_{in,g}} = f^N, \quad f = \sqrt[N]{F}, \tag{3.28}$$

$$\text{In DML: } \frac{S_{N+1}}{S_1} = A^{\frac{N}{2}}, \quad F_{DML} = \frac{S_{N+1}}{S_1} = f_{DML}^{\frac{N}{2}}, \quad A = f_{DML} = \sqrt[\frac{N}{2}]{F_{DML}}, \tag{3.29}$$

where, assuming $S_1 = 1$, S_{N+1} can be extracted from:

$$\frac{(S_{N+1}+1)WL_{min}}{(S_1+1)WL_{min}} = \frac{S_{N+1}+1}{2} = \frac{C_{Load}}{C_{in,g}}. \tag{3.30}$$

This methodology supports the calculation of sizing factors for a given N size chain of inverters. Next, it will be extended to derive the optimal chain length N_{opt} under a given load capacitance. The problem will be defined as derived by Sutherland et al. [1–3].

Consider a path of DML gates containing n_1 stages, to which we append n_2 additional DML inverters to obtain a path with $N = n_1 + n_2$ stages. We assume that the original n_1 stages cannot be altered, except for scaling, since they perform necessary logic functions. However, any positive n_2 can be chosen to reduce the delay of the chain. Moreover, assuming that the optimum length will be larger than n_1, we also assume that n_2 is even and that the logic function will not be altered. It has been shown that adding CMOS inverters to a path does not affect the electrical effort of the path. Since the sizing of DML inverters is very different from CMOS and they are added in the subsequent *Type-A/Type-B* structures, the electrical effort of the path is changed by adding these DML buffers. By using the term for the definition of LE, it can be demonstrated that the LE of the DML A–B inverter pair (for large $i's$) can be approximated by: $LE_{invA} \cdot LE_{invB}\big|_{LARGE\,i's} \cong \frac{1}{2} \cdot \frac{\mu_{n/p}}{2}$.

The electrical effort factorization means that for $\mu_{n/p} = 4$ alone, the electrical effort of the chain will not be affected. The solution can be presented for any value of $\mu_{n/p}$; however, in this chapter we assume $\mu_{n/p} = 4$, for simplicity. Note that this approximation is valid for many nanometer processes [10]. The delay of the whole chain is represented by the sum of the delays of all n_1 logic stages and all the added n_2 inverters. Differentiating the chain delay by N and equating to 0 yields:

$$\underbrace{\gamma(\mu_{n/p}^{-0.5} + \mu_{n/p}^{0.5})}_{C_1} + \sqrt[N]{F_{DML}} - \frac{\sqrt[N]{F_{DML}} * \ln(F_{DML})}{N} = 0. \qquad (3.31)$$

where $C_1 = \gamma(\mu_{n/p}^{-0.5} + \mu_{n/p}^{0.5})$ and the optimal sizing factor can be numerically solved using: $f_{DML\_opt} = \exp(1 + C_1/f_{DML\_opt})$, which leads to $f_{DML\_opt} = 4.65$. As developed in Sect. 3.2, for large F_{DML}, which means that $N_{opt} > N_{minimum}$, N_{opt} can be approximated by:

$$N_{opt} \cong \log_{f_{DML\_opt}}(F_{DML}). \qquad (3.32)$$

As presented by Sutherland et al. [1–3], the deviation from the minimum delay obtained under N_{opt} length implementation can be expressed as:

$$\frac{D_{opt}(Dev \cdot N)}{D_{opt}(N_{opt})} = \frac{C_1 \cdot Dev + Dev \cdot f_{opt}^{1/Dev}}{C_1 + f_{opt}}, \qquad (3.33)$$

where Dev represents the deviation factor from N_{opt}, the lowercase opt stands for optimal, $D_{opt}(N)$ is the optimal delay for a given general N, and $D_{opt}(N_{opt})$ is the optimal delay with the optimal N.

Figure 3.3 depicts the normalized delay as a function of the normalized $N = N_{opt}$ for the CA method (from Eq. (3.33)).

The graph values were normalized to $D_{opt}(N_{opt})$. Note that the only difference as compared to the CMOS solution is the constant C_1 which affects the graph slopes. As depicted in the figure, DML behaves similarly to the CMOS solution, and an overestimation in N is preferable than its underestimation in terms of delay. For example, when $N = N_{opt}/2$, a 68.8% deviation in delay is observed. However, when $N = 2N_{opt}$, the deviation drops to 30.2%.

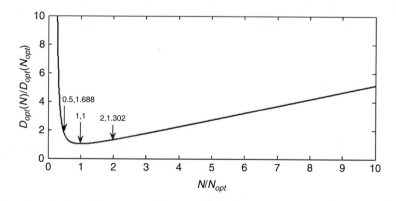

Fig. 3.3 Normalized delay as a function of normalized N/N_{opt} for the CA method

3.3.5 The Semi-approximated Method (SA) for DML Sizing Factors of an Inverter Chain

As a compromise to the methods presented in the previous subsections, a semi-approximated (SA) approach was developed. The SA approach achieves relatively high precision with reduced computational effort with respect to the CS method. This is done by only neglecting the first and the second terms of Eq. (3.20), as compared to omitting all the terms of the gate and drain capacitances (CA method). The solution to the SA is very simple, but in addition to the standard CMOS-LE optimization calculation, the designer needs to use a simple lookup table (given in advance). Since this method is beyond the scope of this chapter, it only appears in Appendix A.

3.4 Generalizing the DML-LE Method for Complex Gates and Branches

This subsection presents a generalization of the DML-LE method for complex gate topologies and intersecting nets and justifies the efficiency of these methods for complex DML gates. It is shown that the CA and SA methods achieve high precision results. The CA method is generalized below. Note that by implementing the same approach, the SA method can also be generalized.

3.4.1 Exploring a General DML Gate Delay Structure

Equation (3.20) described the general delay expression for a DML gate. This expression can be generalized by taking into account the branching effort and the mobility ratio factor which differentiates *Type-A* from *Type-B* gates:

$$D = t_{p0\_DML} \left(\overbrace{\frac{(N_{S\_drain} s_i + N_{\min\_drain})}{2 s_i}}^{p\_DML} \gamma + \underbrace{\underbrace{\overbrace{\frac{(X_{S\_gate} s_i + 1)}{2 \cdot s_i}}^{LE\_DML} \cdot \underbrace{b_i}_{b\_DML}}_{(f \cdot b)\_DML} \cdot \frac{C_{Load\_on}}{(X_{S\_gate} s_i + 1)}}_{EF\_DML} \right),$$

(3.34)

where C_{Load_on} denotes the load capacitance on the critical path under optimization, $N_{\min\_drain}$ is the number of complementary network transistors of gates directly connected to the output (in terms of minimal width transistors), N_{S_drain} is the number of transistors in the evaluation network and the precharge transistors that are directly connected to the gate output (these transistors are sized with the S factor), and X_{s_gate} is the upsizing factor of the transistors in the evaluation network of a gate, which expresses the upsizing of these transistors relative to S (similarly to X_{s_Load}, if we represent the load capacitance in terms of gate capacitance). For a *Type-B* gate the expression on the right side of Eq. (3.35) needs to be multiplied by $\mu_{n/p}$. Table 3.2 lists the delay expressions for the approximated and non-approximated stages with and without branching. The table also lists the approximated branching effort version. The branching effort in CMOS Eq. (3.20) is independent of the sizing factors if all the gates in the same stage i are sized by the same S_i. Since in DML the branching effort depends on the sizing factors S_i, it should be approximated to simplify the delay expression. The approximated branching effort expression appears in Eq. (3.35), where $\sum X_{S\_gate}$ stands for summing all the X_{s_gate} of levels i and $X_{s\_gate\_on\_path\_i}$ is the only X_{s_gate} factor which is on the optimized path.

$$b_{i\_Approximated} = \frac{\displaystyle\sum_{(On+Off)\ Path\_i} X_{S\_gate}}{X_{S\_gate\_on\_path\_i}} \tag{3.35}$$

In Sect. 3.3.4 two expressions for LE and p were approximated for inverters: $2s_i + 1 \rightarrow 2s_i$, $s_{i+1} + 1 \rightarrow s_{i+1}$, where S_i is the i'th stage sizing factor (for $i > 2$). In this subsection, it is shown that in most cases of complex gates, the approximated values are even more negligible. Table 3.3 lists several approximation examples for p and LE for several complex gates.

As listed in Table 3.3, the approximation error decreases when implementing complex gates (compared to an inverter) in the majority of cases. As discussed above, the preferred DML gate topology is such that the precharge transistor is placed in parallel to the stacked transistors [11–13], i.e., NOR in *Type-A* is preferred over a NAND, and NAND in *Type-B* is preferred over NOR. Therefore, Table 3.3 only lists the preferred gates.

3.4.2 Delay Optimization Under the Complete Approximated (CA) Model for Complex Gates

Optimizing the delay under the CA model requires the use of the approximated LE, P for all stages of the design along with the approximated branching shown in Table 3.2. P, LE, and f are marked with a lowercase "approximated" to emphasize that they are approximated. Summing Eq. (3.35) listed in Table 3.2 for the length N chain (*Type-A* and *Type-B* alternately), then differentiating by the sizing factors and equating to 0, results in:

Table 3.2 Generalized DML delay expressions for "type A" gates (for "type B" gates all equations need to be multiplied by $\mu_{n/p}$)

Without branching	$$t_{p0\_DML}\left(\underbrace{\gamma + \underbrace{\dfrac{N_{S\_drain}}{2}}_{p\_DML\_Approximated}\;\underbrace{\dfrac{X_{S\_gate}}{2}}_{LE\_DML\_Approximated}\;\dfrac{X_{S\_Load}\cdot s_{i+1}}{X_{S\_gate}\cdot s_i}}_{(f\text{-}b)\_DML\_Approximated}\right)$$
Including branching	$$t_{p0\_DML}\left(\underbrace{\gamma + \underbrace{\dfrac{N_{S\_drain}}{2}}_{p\_DML\_Approximated}\;\underbrace{\dfrac{X_{S\_gate}}{2}}_{LE\_DML\_Approximated}\;\underbrace{\dfrac{\sum\limits_{On+Off\,Path\_i}(X_{S\_gate})}{X_{S\_gate\_on\_path\_i}}\,\dfrac{X_{S\_Load\_on}\cdot s_{i+1}}{X_{S\_gate}\cdot s_i}}_{b\_DML\_Approximated}}_{(f\text{-}b)\_DML\_Approximated}\right)}_{EF\_Approximated}$$

Non-approximated stages—For $i = 1, 2$

Without branching	$$t_{p0\_DML}\left(\underbrace{\gamma + \underbrace{\dfrac{(N_{S\_drain}s_i + N_{min\_drain})}{2s_i}}_{p\_DML\_non\text{-}Approximated}\;\underbrace{\dfrac{(X_{S\_gate}s_i + 1)}{2\cdot s_i}}_{LE\_DML\_non\text{-}Approximated}\;\dfrac{C_{Load}}{(X_{S\_gate}s_i + 1)}}_{(f\text{-}b)\_DML\_non\text{-}Approximated}\right)$$
Including branching	$$t_{p0\_DML}\left(\underbrace{\gamma + \underbrace{\dfrac{(N_{S\_drain}s_i + N_{min\_drain})}{2s_i}}_{p\_DML\_non\text{-}Approximated}\;\underbrace{\dfrac{(X_{S\_gate}s_i + 1)}{2\cdot s_i}}_{LE\_DML\_non\text{-}Approximated}\;\underbrace{b_i\,\dfrac{C_{Load\_on}}{(X_{S\_gate}s_i + 1)}}_{b\_DML\_non\text{-}Approximated}}_{(f\text{-}b)\_DML\_non\text{-}Approximated}\right)}_{EF\_non\text{-}Approximated}$$

Table 3.3 P and LE approximations for several complex gates

Gates	p and LE approximations
NAND3_B, NOR3_A	$(4s_i + 1)/s_i \rightarrow 4$
OAI21_B, AOI21_A	$(4s_i + 2)/s_i \rightarrow 4$
OAI21_A, AOI21_B	$(5s_i + 2)/s_i \rightarrow 5$
NAND2_B, NOR2_A	$(3s_i + 1)/s_i \rightarrow 3$

Table 3.4 Complex gate chain sizing factors (S_i) for the CA method

S_1	S_2	S_3	S_4	S_{N+1}
1	$EF \frac{\sqrt{\mu_{n/p}}}{\langle 1 \rangle X_2}$	$EF^2 \frac{1}{\langle 1 \rangle \langle 2 \rangle X_2 X_3}$	$EF^3 \frac{\sqrt{\mu_{n/p}}}{\langle 1 \rangle \langle 2 \rangle \langle 3 \rangle X_2 X_3 X_4}$	$EF^N \frac{1}{\langle 1 \rangle \langle 2 \rangle \langle 3 \rangle ... \langle N \rangle X_2 X_3 ... X_{N+1}}$

$$\text{odd } i : \quad \frac{S_i}{S_{i-1}} = \frac{S_{i+1}}{S_i} \frac{\langle i \rangle X_{i+1}}{\mu_{n/p} \langle i-1 \rangle X_i}, \tag{3.36}$$

$$\text{even } i : \quad \frac{S_i}{S_{i-1}} = \frac{S_{i+1}}{S_i} \mu_{n/p} \frac{\langle i \rangle X_{i+1}}{\langle i-1 \rangle X_i}, \tag{3.37}$$

where, for simplicity, the next terms are defined:

$$\langle i \rangle = \frac{LE_{DML\_i} b_{DML\_i}}{X_{s\_gate\_i}} ; X_{s\_gate\_i} = X_i. \tag{3.38}$$

The sizing factors series, S_i, are presented in Table 3.4 and were extracted from the following equation:

$$EF_{DML} = \sqrt[N]{\frac{S_{N+1} X_{N+1}}{S_1 X_1} \prod_{i=1}^{N} LE_{DML\_i} \prod_{i=1}^{N} b_{DML\_i}}, \tag{3.39}$$

where $F_{DML} = \frac{S_{N+1} X_{N+1}}{S_1 X_1}$ and the branching and logic effort values are approximated. For the general case, the sizing factors can be calculated in a series from the last $i = N + 1$ (load) to the first $i = 1$ by using the following expressions:

$$\begin{aligned} \text{even } i : & \quad S_{i+1} X_{i+1} = \frac{EF_i}{LE_i \cdot b_i \cdot \sqrt{\mu_{n/p}}} S_i X_i \\ \text{odd } i : & \quad S_{i+1} X_{i+1} = \frac{EF_i \cdot \sqrt{\mu_{n/p}}}{LE_i \cdot b_i} S_i X_i \end{aligned}. \tag{3.40}$$

The optimal number of stages for a given load can be computed similarly to Eq. (3.31):

$$\underbrace{\gamma (\mu_{n/p}^{-0.5} + \mu_{n/p}^{0.5})}_{C_1} + \sqrt[N]{PE_{DML}} - \frac{\sqrt[N]{PE_{DML}} * \ln(PE_{DML})}{N} = 0, \tag{3.41}$$

where the optimal sizing factor can be numerically solved using: $EF_{DML\_OPT} = e^{\left(1 + \frac{c_1}{EF_{DML\_OPT}}\right)}$ which leads to $EF_{DML_opt} = 4.65$. Similar to the results presented above for inverters, for large F_{DML}, N_{opt} can be approximated by:

$$N_{opt} \cong \log_{EF_{opt\_DML}}(PE) = \log_{EF_{opt\_DML}}(F * LE * B). \qquad (3.42)$$

Thus, the CA computation effort and the final mathematical results for complex gates are intuitive and as user-friendly as for the standard CMOS-LE. The SA method, which can also easily be derived for the complex gates, is also very simple and uses an additional lookup table.

3.5 Comparing the DML-LE Methods

This subsection presents a comparison of the SA, CS, and CA techniques in terms of their simplicity, accuracy, and dependence on the optimal number of stages and how these affect the delay.

3.5.1 Delay Error for a Given N

In order to compare the solutions in terms of error in delay, the CS method, which is the most precise, was chosen as the reference. Delays for the same chains ($N = 6$) were calculated using expressions presented in the previous subsections. Typical $\gamma = 1$ and $\mu_{n/p} = 4$ values [10] were assumed. We first examine the sizing factor deviation for the SA, CS, and CA techniques, as shown in Fig. 3.4. Two extreme load cases are considered. Figure 3.4a, which depicts the case of a large load of $C_{Load} = 50 \cdot C_{in}$, presents sizing factor deviations of less than -35% for both approximated methods. Crucially, these negative deviations result in smaller transistor sizes and smaller areas in the approximated methods. This improvement in area achieved by the CS and SA methods comes at the expense of a slight increase in the delay, as shown in the figures.

Figure 3.4b shows that for a small load of $C_{Load} = 5C_{in}$, both approximated methods present relatively large negative deviations in sizing factors (max -56% for SA and -67% for CS) which means that they are more area efficient for small loads. Note that the realistic maximum load capacitances in standard logic chains and global interconnect (with repeaters) are around $10C_{in} - -20C_{in}$. However, recently, a unified logical effort that utilizes and optimizes the sizing of logic gates as repeaters in long interconnect wires was proposed [14]. It showed that a load capacitance of around $100C_{in}$ was realistic. Since the goal of this book is not to limit the DML-LE to standard cases, we provide results for loads of up to $100C_{in}$. The delay for all methods as a function of the load capacitances is presented in

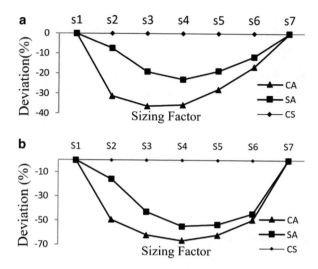

Fig. 3.4 Deviation of sizing factors from the complete solution: (**a**) $C_{Load} = 50C_{in}$, (**b**) $C_{Load} = 5C_{in}$

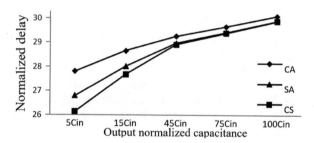

Fig. 3.5 Normalized delay as a function of the normalized C_{LOAD}

Fig. 3.5. These results were normalized to the delay of a minimum-sized *Type-A* DML inverter. This was done by calculating the expression in Eq. (3.20) with the sizing factors for each method. The graph shows that as C_{Load} increases, the relative error decreases significantly. For example, for the SA technique, minimum and maximum errors of 2.49% and 0.01% were achieved for loads of $C_{Load} = 5C_{in}$ and $C_{Load} = 100C_{in}$, respectively. For the CA method, these errors increased to 6.37% and 0.19%. Note that the choice of best solution depends to a great extent on the level of accuracy required. In most cases, it is safe to say that due to the error inherently related to the LE method, the SA method is sufficient with a very small maximal error of $\sim 2\%$. As can be seen, both methods yield highly precise results. It is clear that for cases with small load capacitances and/or short chains, the SA method is preferable. However, the CA method is sufficient for most cases.

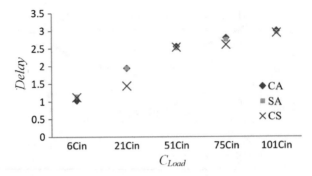

Fig. 3.6 N_{opt} for all three methods and for different C_{LOAD}

3.5.2 N_{opt} *Comparison*

N_{opt} was calculated for various output load capacitances, for all three methods. The N_{opt} calculation for the CS method requires solving a set of $N-1$ nonlinear equations (see Eq. (3.23)). $MATLAB$ software was used to numerically solve the equations by guessing roughly the correct solution area for each N until the numerical simulations converged to the correct solution. In contrast to the CS method, the N_{opt} calculations using the CA and SA techniques were simple and straightforward, as presented in the previous subsection.

Figure 3.6 compares N_{opt} as a function of C_{Load} for all methods. It is clear that the deviations in N_{opt} are quite small for all methods. As discussed in the next subsection, in most cases the error in N_{opt} leads to a zero error in the delay, since only integer Ns are possible.

3.5.3   *Delay Error for a Variable N*

In Sect. 3.3.4, the normalized delay as a function of the normalized N to N_{opt} was formulated for the CA method; see Fig. 3.3. In this subsection, the same ratio is derived for the CS method, as shown in Fig. 3.7. As expected, the deviation in delay depends on the load capacitance. Similar to the behavior presented in Fig. 3.3, an overestimation of N is preferable to its underestimation. Moreover, for different loads, the curve *rotates* around the normalization point. Therefore, overestimating N is even more advantageous for larger loads.

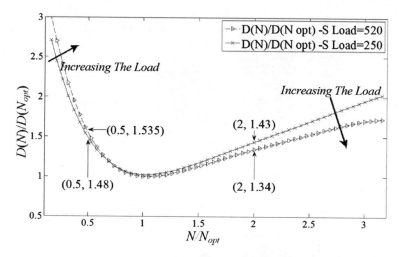

Fig. 3.7 Normalized delay of the CS method as a function of normalized N/N_{opt} for different loads

3.6   Example of a DML-LE Evaluation: a 40 nm Process

Here we evaluate the methodologies by comparing the LE optimization results to those derived by the Cadence Virtuoso optimizer tool. We evaluated on two different logic networks, implemented in a low-power standard 40 nm technology. First, a simple logic chain, identical to the chain theoretically analyzed in Sect. 3.3.2, with $N = 6$ is discussed. The objective of this test was to compare the results of the delay optimization of all the LE methods to the results of the Cadence Virtuoso optimizer tool for different loads. The simulated delay (SPICE) of the chain, sized according to the DML-LE CS method, and the delay optimization derived by the Cadence optimizer are shown in Fig. 3.8.

In addition, to compare the precision of the proposed DML-LE methodology to a standard CMOS-LE, the same testbench was constructed with CMOS logic and was optimized with the Cadence Virtuoso optimizing tool. The results of this test are presented in Fig. 3.9.

The analysis indicates that both cases achieved a very similar maximum error (3% for CMOS-LE and 2.4% for DML-LE). The maximum LE error in comparison to the optimal optimizer solution was 3% for CMOS-LE. It is worth noting that a variety of different chains were examined and very similar errors were observed for all test cases. Figure 3.10 compares the SA, CS, and CA methods as well as the results of the Cadence Virtuoso optimizer. As visualized in the figure, all the methods presented very similar results, which implies that the CA methodology should be preferred over the CS and SA due to its significantly less computation complexity. Note as well that the DML-LE sizing factors of the CA and SA techniques were highly similar to those of the optimizer (errors of 2–3% or less).

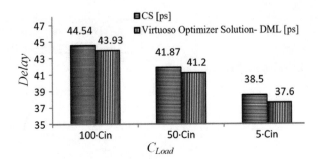

Fig. 3.8 Six DML inverter chain delays as a function of the normalized load capacitance for the DML CS method and for the Cadence optimizer solution

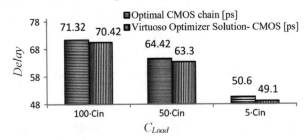

Fig. 3.9 Six CMOS inverter chain delays as a function of the normalized load capacitance for the LE solution and for the Cadence optimizer solution

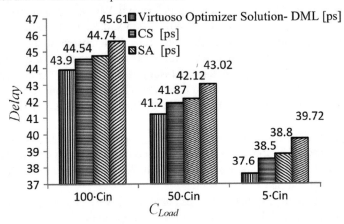

Fig. 3.10 Six DML inverter chain delays as a function of the normalized load capacitance of all the DML-LE solutions and for the Cadence optimizer solution

Thus, the area and energy overhead were extremely low, as compared to the optimizer solution.

To investigate the performance of the SA, CS, and CA techniques in depth, the deviations in delays from the complete solution which is optimal for DML were calculated, as shown in Fig. 3.11.

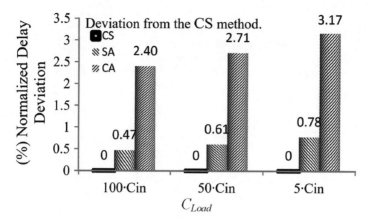

Fig. 3.11 DML CA and SA delay deviations from the CS method

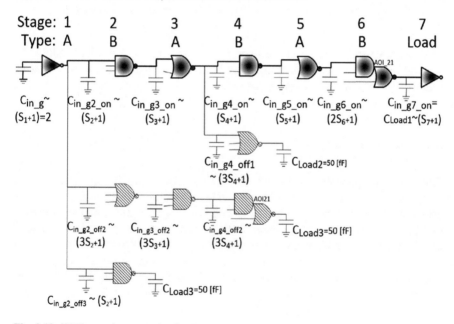

Fig. 3.12 DML complex network scheme

As expected, CA precision improved for larger loads. The precision of the SA was also very high for cases of small loads (large N). Therefore, the SA method provides a good tradeoff between computational complexity and precision for these cases. To evaluate the performance of an optimized DML complex logic network which includes branches (schematically shown in Fig. 3.12), it was compared to the same CMOS network. While the CMOS network was optimized using a standard LE, the DML network was sized according to the SA DML-LE methodology. Figure 3.13 shows the results of this comparison. Application of the DML-LE effort

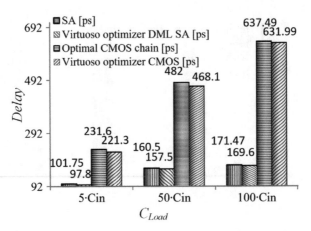

Fig. 3.13 DML and CMOS complex network delays, optimized using DML- and CMOS-LEs, respectively. In addition, the Virtuoso optimizer results for both methods are presented

in a complex network resulted in a maximum error of only ~4.5% for small loads, as compared to the Cadence optimizer results. This is very close to the ~3.8% error achieved by the CMOS-LE optimization.

3.7   Conclusion

The DML-LE approach enables efficient optimization of DML logic networks for maximum performance in the dynamic mode of operation. DML logic, optimized according to LE methods, provides extended flexibility in optimizing the structures of DML networks. This optimization utilizes DML's inherent properties of significantly reduced parasitic capacitance and ultra-low-power dissipation in the static operation mode [11–13]. Three different approaches that trade off computation complexity and accuracy were discussed in this chapter. The complex CS method was only examined to frame the error analysis comparison for the other methods. The CA method is identical to the CMOS-LE computation, with very small error, and the SA method is also identical to the CMOS-LE computation with the addition of one more lookup table (which is easily derived for all cases and loads). With these tools in hand, a design can achieve very high performance results. For designers, this chapter should shed light on the advantages and tradeoffs of each method. The simulation results, which were carried out in a standard 40 nm process, document the efficiency of the approach as compared to CMOS-LE.

References

1. R.F. Sproull, I.E. Sutherland, Logical effort: Designing for speed on the back of an envelope. IEEE Adv. Res. VLSI, 1–16 (1991)
2. I.E. Sutherland, R.F. Sproull, D.F. Harris, *Logical Effort: Designing Fast CMOS Circuits* (Morgan Kaufmann, Boston, 1999)
3. A. Kabbani, D. Al-Khalili, A.J. Al-Khalili, Delay analysis of cmos gates using modified logical effort model. IEEE Trans. Comput. Aided Des. Integr. Circ. Syst. **24**(6), 937–947 (2005)
4. B. Lasbouygues, S. Engels, R. Wilson, P. Maurine, N. Azémard, D. Auvergne, Logical effort model extension to propagation delay representation. IEEE Trans. Comput. Aided Des. Integr. Circ. Syst. **25**(9), 1677–1684 (2006)
5. S.K. Karandikar, S.S. Sapatnekar, Technology mapping using logical effort for solving the load-distribution problem. IEEE Trans. Comput. Aided Des. Integr. Circ. Syst. **27**(1), 45–58 (2008)
6. S.K. Karandikar, S.S. Sapatnekar, Logical effort based technology mapping, in *Proceedings of the 2004 IEEE/ACM International Conference on Computer-aided Design* (IEEE Computer Society, Washington, 2004), pp. 419–422
7. P. Rezvani, A.H. Ajami, M. Pedram, H. Savoj, Leopard: a logical effort-based fanout optimizer for area and delay, in *Proceedings of the 1999 IEEE/ACM International Conference on Computer-aided Design* (IEEE Press, Piscataway, 1999), pp. 516–519
8. J.M. Rabaey, A.P. Chandrakasan, B. Nikolic, *Digital Integrated Circuits*, vol. 2 (Prentice Hall, Englewood Cliffs, 2002)
9. N.F. Goncalves, H. De Man, Nora: A racefree dynamic cmos technique for pipelined logic structures. IEEE J. Solid-State Circuits **18**(3), 261–266 (1983)
10. W.P. Chen, P. Su, J. Wang, C. Lien, C. Chang, K. Goto, C. Diaz, A new series resistance and mobility extraction method by bsim model for nano-scale mosfets, in *2006 International Symposium on VLSI Technology, Systems, and Applications* (2006)
11. A. Kaizerman, S. Fisher, A. Fish, Subthreshold dual mode logic. IEEE Trans. Very Large Scale Integr. (VLSI) Syst. **21**(5), 979–983 (2013)
12. I. Levi, O. Bass, A. Kaizerman, A. Belenky, A. Fish, High speed dual mode logic carry look ahead adder, in *2012 IEEE International Symposium on Circuits and Systems* (IEEE, Piscataway, 2012), pp. 3037–3040
13. I. Levi, A. Kaizerman, A. Fish, Low voltage dual mode logic: Model analysis and parameter extraction. Microelectron. J. **44**(6), 553–560 (2013)
14. A. Morgenshtein, E.G. Friedman, R. Ginosar, A. Kolodny, Unified logical effort—a method for delay evaluation and minimization in logic paths with interconnect. IEEE Trans. Very Large Scale Integr. (VLSI) Syst. **18**(5), 689–696 (2010)

Chapter 4
Low-Voltage DML

This chapter examines DML performance, energy consumption, static noise margins, delay distribution, robustness, and other design metrics under low-voltage operation. It still focuses on the gate level and DML operations in subthreshold and near-threshold regions illustrated using the transregional model. Measurement results for fabricated test structures of a variety of DML benchmarks are presented, covering a wide range of operating conditions, supply voltages, etc. Whereas the last chapter mainly dealt with performance optimization of the DML dynamic mode, the main goal of this chapter is to provide an in-depth description of the superior performance of DML with regard to robustness and process variation immunity, and its operation at low voltages. The DML designs are compared to standard CMOS and (dynamic) domino to provide the reader with a better grasp of DML's key features, compared to current alternatives.

4.1 Introduction

The progress achieved in the fields of portable, wearable, and internet-of-things devices, their ever-increasing popularity (and projected market growth), and the fact that battery technology cannot keep up the pace have prompted researchers to actively explore the field of low-power design [1–6]. Low-voltage digital circuit design in particular is a very popular approach for ultra-low-power applications [7–9]. Typically, the circuits operate in the subthreshold (ST) and/or near-threshold (NT) regions, from a supply voltage (V_{DD}) that is close or even less than the threshold voltages of the transistors. This aggressive reduction of supply voltage leads to significant savings in dynamic and static power. Unfortunately, low-voltage circuits suffer from degradation in performance, increased sensitivity to process variations, and device mismatch [10]. Extensive research has attempted to overcome

© Springer Nature Switzerland AG 2021
I. Levi, A. Fish, *Dual Mode Logic*, https://doi.org/10.1007/978-3-030-40786-5_4

these sensitivities under low-voltage regimes, including suggestions such as the adaptive body bias or dynamic voltage scaling for adaptive power applications [11–16].

As presented in Chap. 1, for decades, CMOS has been considered the most efficient VLSI design methodology. CMOS gates are very robust, achieve rail-to-rail logic levels, have strong on and off states, and until the advent of recent processes also had exceptionally low static power consumption. Unsurprisingly, CMOS logic has also become the most common design logic family for low-voltage operation. In many cases, it achieves more robust operation than its Pass Transistor Logic (PTL) and dynamic logic counterparts. However, the low-voltage CMOS delay is significantly degraded, making ST and NT CMOS designs impractical for many applications (as briefly discussed in Sect. 1.2.1).

Low-voltage dynamic logic, such as domino, has been proposed as a better alternative with higher performance than low-voltage CMOS designs [17]. However, the challenges of dynamic logic such as charge sharing, susceptibility to glitches and crosstalk noise, and sensitivity to process variations in nanoscaled technologies make it a poor choice.

By contrast, DML logic allows both ultra-low power dissipation and high performance under extreme low-voltage operation. In fact, the unique structure of the DML gate, which inherently contains CMOS complementary networks, leads to robust operation of the DML gate in the dynamic mode even under low voltages. In addition, the ability to operate as a dynamic gate compensates for the lack of performance of the CMOS part of the DML gate. This chapter reports simulations and measurements showing that DML gates are fully functional at supply voltages as low as 0.3 V. In low-power 40 nm technology, the chains of subthreshold basic DML-NAND/NOR/FA gates achieve a tenfold improvement in speed in the dynamic mode compared to the standard CMOS while dissipating only 150% more power. In the static mode, it achieves a 50% reduction of power dissipation, as compared to basic domino logic, at the expense of a single magnitude decrement in performance. Recall that switching between modes is very easy and can be performed on-the-fly, making DML ideal for computing platforms where a flexible workload is required. The energy consumption of the DML static mode leads to less energy consumption and lower Minimum Energy Point (MEP), as compared to conventional CMOS operation. These metrics are the hallmarks of the flexibility of the DML family in terms of achieving high-performance or low-power operation, as a function of time-dependent system requirements.

4.2 DML Under Low-Voltage Operation

The optimization space of DML gates, like the majority of VLSI designs, is composed of area, power, and speed. Since ST and NT designs suffer from reduced performance, this chapter targets DML gate optimization for speed. To optimize the performance of DML gates, we use the logical effort (LE) technique to evaluate the delay [18] as was done in the previous chapter. Recall that in the LE technique, the gate delay (d) can be expressed as the sum of the stage effort (f) and the parasitic capacitance (p):

$$d = f + p, \tag{4.1}$$

where $f = g \cdot h \cdot b$, g is the logical effort of the stage,[1] h is the electrical effort, and b is the branching effort. The main goal of this chapter is to show the feasibility of using LE for less trivial transistor current models. To do so, we first need to evaluate the logical effort (g) of the gate, which is defined as the ratio of the input capacitance of the gate to that of an inverter, assuming that both gates drive the same current. g is an intrinsic property of the gate and is constant. In order to evaluate g, we need to set the transistor widths so the DML gate can deliver the same amount of output current as the inverter. Section 4.3 presents a transregional current model for low-voltage operation, i.e., ST and NT regions. This model is used to evaluate the capabilities of the stacked transistor fitting parameters to achieve the same current as driven by a single transistor. Based on these derived fitting parameters, the DML sizing architecture is presented and the logical effort parameters are calculated.

4.3 DML Modeling and Sizing Using the Transregional Model

4.3.1 Modeling I_{on} Using the Transregional Model

In 2006, Keane et al. [19] presented a subthreshold logical effort (LE) framework that had unconventional device sizing. Given the simplicity of their approach and modeling, we apply this to DML gates and extend its application to include the NT region, which complies with advanced nanoscale processes. Since the characteristics of MOS transistors operating in the ST and NT regions are substantially different from transistors operating in strong inversion, we use the transregional model developed by Harris et al. [20]. In this model, the on-current (I_{on}) of the transistor is modeled by:

$$I_{on} = I_0 W e^{\frac{V_{DT} - \alpha V_{DT}^2}{n v_T}}. \tag{4.2}$$

where V_{DT} stands for V_{DD}-V_T, α and n are empirical fitting parameters, and v_T is the thermal voltage. For purposes of illustration, the model parameters were derived by curve-fitting Spectre simulations for the chosen low-power 40 nm technology.

---

[1] g was denoted by LE in the previous chapter. In this chapter, for conciseness, we use the original notations.

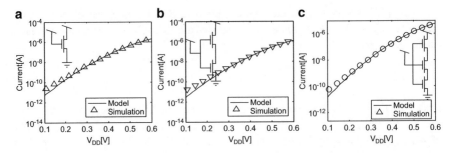

Fig. 4.1 I_{ON} current *vs*. V_{DD}: theoretical calculation for the transregional model *vs*. the simulation results. (**a**) Single transistor fitting, (**b**) two-stack transistor fitting, and (**c**) three-stack transistor fitting

I_{on} is a function of V_{DT}, so that changes in V_T caused by process variations or body biasing do not require the re-fitting of I_0, α and n. V_{DD} was swept to extract the model parameters for Eq. (4.2), as shown in Fig. 4.1a. The figure plots the simulation results against the calculated model. The model must be accurate in the functional ST and NT regions to be further used to derive the required transistor widths. As shown in Fig. 4.1a, throughout the entire range from the functional subthreshold region ($V_{DD} > 0.25$ V) to the NT vicinity, the least square error fit had an average error of less than 1% and a maximum error of 9%.

To show that the transregional model is not restricted to modeling of a single transistor, it was also examined for a stack of two and three transistors. Figure 4.1b and c shows that the model fits the operational regions accurately with an average least square fit error of 3% and a maximal error of 14%. The modeling of stacks of several transistors is important during LE development for DML. Note that the accuracy is very high because the transregional model is "curve fitted." The maximum error of 9–14% is achieved for 250 mV supply voltage. For supply voltages exceeding 250 mV, the error falls off very rapidly and becomes negligible.

4.3.2   Low-Voltage DML Sizing Methodology

In the dynamic mode, a fast evaluation period is critical. For this reason we analyze a DML topology where the precharge transistor is placed in parallel to the stacked transistors (i.e., NORs in *Type-A* and NANDs in *Type-B*). In addition, we analyze the size of the footer and evaluation transistors (i.e., a stack of two transistors in an optimal parallel evaluation net). The vast majority of gates are un-footed. In complex logic gates (i.e., AOI/IOA cells) the evaluation net can comprise more than one transistor, even without a footer. The complementary serial transistors, which are parallel to the precharge transistor, are sized to minimal width to decrease the gate capacitances and intrinsic delay and thus enable fast dynamic operation. The precharge transistor also needs to have minimal width, to limit leakage currents.

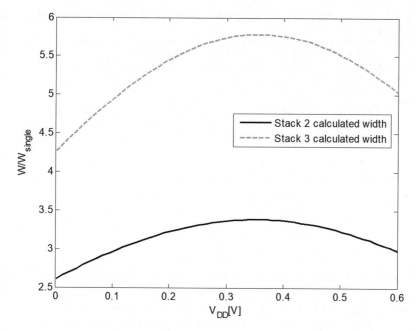

Fig. 4.2 Ratio of the calculated widths of transistors in a stack to the width of a single transistor

The precharge transistor can be sized even larger, as was described in the previous chapter. The gain in output capacitance from using a minimally sized precharge transistor is negligible.

We implement the transregional current model to calculate the widths of the footer and evaluation transistors (W') required to drive the same on-current as drives a single transistor (W). The $I_{on,single}$ of a single transistor is equated to $I_{on,2\,stack}$ and $I_{on,3\,stack}$, and W is extracted as a function of the fitting parameters, as can be seen in Eq. (4.3). The variables marked with a tick ("'") are for the stacked transistors.

$$\frac{W'}{W} = \frac{I_0}{I_0'} \cdot \exp(\frac{V_{DT} - \alpha V^2{}_{DT}}{nv_T} - \frac{V'_{DT} - \alpha'(V'_{DT})^2}{n'v_T}). \qquad (4.3)$$

Figure 4.2 shows the ratio of the calculated widths of the transistors in a stack to the width of a single transistor. Note that the widths of the transistors in a stack are not constant, but rather a function of V_{DD}. Thus, the optimal width varies with the supply voltage and is different from region to region. This optimization is advantageous for ST and NT region circuits, since traditional sizing (a ratio of two and three, for the two and three transistor stacks, respectively) is not precise at low voltages. For example, at a 0.3 V supply voltage, the sizing factor needed to obtain the same current through two-stack transistors as for the case of a single minimal transistor is 3.3 (for a three-stack, 5.6, and for a four-stack, 10). Nevertheless, for the strong inversion regime, the width converges to nominal values (not shown in Fig. 4.2).

Table 4.1 Optimal transistor width for CMOS, dynamic logic, and DML in all topologies for $V_{DD} = 0.3$ V

Footed DML *Type-A*	Footed DML *Type-B*	Un-footed DML *Type-A*	Un-footed DML *Type-B*	Dynamic	CMOS
NOR3					
NAND3					

Based on this analysis, we calculated the optimal transistor sizing of basic DML gates. Table 4.1 presents an example of optimized transistor sizing, normalized to the minimal transistor width for NAND and NOR gates with a fan-in of 3 for V_{DD} =0.3 V for DML, CMOS, and domino designs for both footed and un-footed topologies. The same stacked-transistor analysis was also used to calculate the widths of CMOS/dynamic transistors. The sizing factor β, which is defined as the optimal ratio for transistors of the pull-up to the pull-down network, was simulated and found to be $\sim$1.5 for the 40 nm process, which is thus used as the factor for all calculations henceforth.

4.3.3   Logical Effort Parameters for Low-Voltage Operation

Using the transistor widths from the previous sections, we calculated the LE parameters shown in Table 4.2. Note that the derived values are smaller than their CMOS/domino counterparts for the un-footed *Type-A* NOR3 and *Type-B* NAND3 gates. These delays are related to DML operation in the dynamic mode. For the static mode, it is clear that DML gates are slower than CMOS because of the unsymmetrical sizing and minimum-sized complementary network. However, due to the reduced input and output capacitances, the degradation in performance in the static mode is not very large. Thus DML designers would clearly benefit from designing their logic in such a way that gates can be constructed with a high-stack pull-up network in *Type-A* and high-stack pull-down network in *Type-B* and avoid as much as possible incorporating other combinations such as un-footed *Type-A* NAND or *Type-B* NOR. This type of design approach yields very fast circuits. Recall that footed gates are used infrequently and are only presented here for completeness.

Table 4.2 Transistor width and LE parameter calculations

Gate	Technology	pMOS width	nMOS width	Clock transistor width	p	g	$d = p + ghb$ for $h=3$ $b=1$
NOR3	CMOS	8.4	1		11.4/3	9.4/3	13.2
	Domino	1	3.3	1	10.9/3	3.3/3	7.2
	Un-footed DML—Type-A	1	1	1	5/3	2/3	3.667
	Un-footed DML—Type-B	8.4	1	1	12.4/3	9.4/3	13.533
	Footed DML—Type-A	1	3.3	1	11.9/3	4.3/3	8.267
	Footed DML—Type-B	15	1	1	19/3	16/3	22.33
NAND3	CMOS	1.5	5.6		10.1/3	7.1/3	10.467
	Domino	1	10	1	11/3	10/3	13.667
	Un-footed DML—Type-A	1	5.6	1	9.6/3	6.6/3	9.8
	Un-footed DML—Type-B	1.5	1	1	6.5/3	2.5/3	4.667
	Footed DML—Type-A	1	10	1	14/3	11/3	15.667
	Footed DML—Type-B	5	1	1	17/3	6/3	11.667

Note: LE for all dynamic logics only takes the evaluated transition into account (i.e., these values are for comparison purposes only in the dynamic mode)

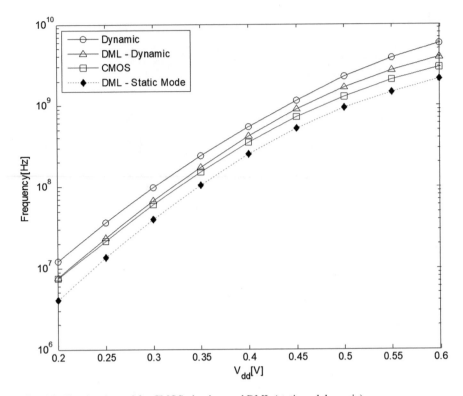

Fig. 4.3 Simulated speed for CMOS, domino, and DML (static and dynamic)

Figure 4.3 shows the frequency of a 40 nm NAND–NOR DML chain (a chain of 20 gates, 10 NAND gates in *Type-B*, and 10 NOR gates in *Type-A*), as well as CMOS and domino chains with a fan-out of 3 of the same length for different supply voltages. It should be noted that a very noticeable gain is achieved when testing more complicated designs, as discussed in the following sections.

A NOR DML gate operating in static mode is on average 33% slower than a CMOS gate. Switching a DML gate from the static mode to the dynamic mode represents an average speed improvement of 2× in the footed topology (e.g., at V_{DD} =0.3 V, whereas the dynamic DML achieves 66 MHz, the CMOS achieves only 50 MHz, and the static DML 35 MHz). In the un-footed topology, an improvement of up to 14× was observed [21]. As expected, domino logic can operate at the highest frequency but is susceptible to process variations. On average, dynamic DML operation consumes 100% more energy than static DML, as discussed in Sect. 4.4.

In this section we derived the logical effort parameters under modeling of the current through a single transistor. An example of a normalized delay by LE analysis based CMOS, domino, and un-footed *Type-A* DML NOR gates with fan-in of 3 is illustrated in Fig. 4.4. The CMOS NOR gate has the largest LE value and hence has

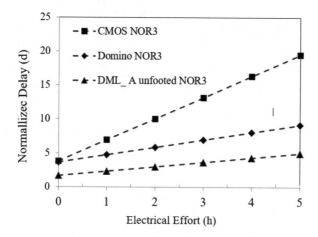

Fig. 4.4 Normalized delay as a function of the electrical effort. The delay is related linearly to *g* and the offset of *p*

the highest delay for any given *h*, whereas the DML un-footed NOR has the smallest LE and parasitic delay, hence the smallest normalized delay for a given *h*. This is consistent with expectations.[2]

4.4 DML Benchmark Measurements

A low-voltage 40 nm DML Full Adder (FA) was designed to illustrate the functionality and robustness of the low-voltage DML methodology. The adder was compared to CMOS and dynamic FAs [22]. The design of the DML FA was based on the optimization developed in the previous subsections such that the sizing of the stacked transistors in all designs was based on the analysis above. Figure 4.5 depicts the implementation of the FA in these logic families. The conventional CMOS (CCMOS) design has 24 transistors, the dynamic design has 20, and the DML has 30 transistors. It should be noted that the increased transistor count in the DML FA does not lead to area increase, since more than 50% are minimum-sized transistors. All the adders were designed and characterized in a standard low-power 40 nm process using the Cadence Virtuoso-based Spectre simulator. Power supplies between 150 and 600 mV were evaluated for energy estimation. Monte-Carlo statistical simulations were conducted at 300 mV to compare the sensitivity of the simulated adders to process variations and mismatch.

---

[2]Note that domino logic in subthreshold voltages (e.g., 0.3 V) is not robust and does not function correctly under variations, as discussed below; therefore, we compare it here to a footed domino implementation to derive the LE parameters.

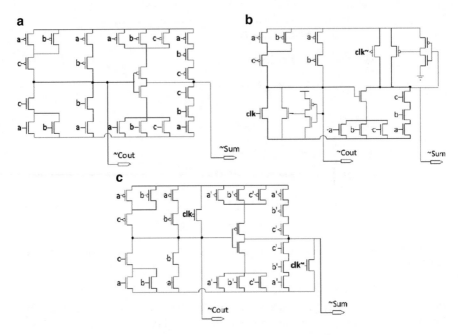

Fig. 4.5 FA implementation: (**a**) CCMOS. (**b**) Dynamic logic (**c**) DML

4.4.1 DML Robustness and Design Metrics Under Low Voltage

To investigate the performance and robustness of the FAs, a Static Noise Margin (SNM) analysis was performed. This test quantitatively estimates the robustness of the design and its susceptibility to process variations. SNM was measured in a very similar way to that introduced by Kwong et al. [23]. This method consists of a butterfly plot analysis and verifies proper V_{OL} and V_{OH} values. The inputs of two FAs are cross-coupled so they act as inverters and feed each other. Then the input voltage is swept, and the two output curves of the FAs are superimposed to create a butterfly curve. SNM is measured as the largest square inscribed in the smaller lobe of the butterfly plot. Since this metric is applicable to static-logic families as is, we used it to evaluate the robustness of CMOS and DML in the static mode. Figure 4.6 shows the Monte-Carlo SNM results. The SNM from static DML was slightly higher than CMOS (86 μ in CMOS as compared to 91 μ in DML). Overall, DML in the static mode appeared to be as robust as CMOS. As implied in Fig. 4.5b, the dynamic version of the FA failed to operate at 300 mV as discussed next.

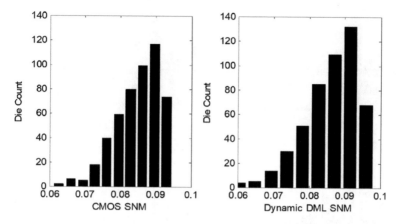

Fig. 4.6 Monte-Carlo SNM analysis, $V_{DD} = 0.3\,\text{V}$. $\mu_{DML\_Static} = 91\,\text{m}$, $\sigma_{DML\_Static} = 8\,\text{m}$. $\mu_{CMOS} = 86\,\mu$, $\sigma_{CMOS} = 6\,\text{m}$

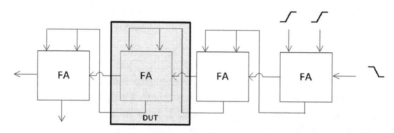

Fig. 4.7 DUT configuration for delay analysis. Inputs are marked with data rising/falling waveforms

4.4.2 Energy and Delay Analysis

Each FA was tested to evaluate energy and delay. The DML gates were tested in both the static and dynamic modes. Each FA implementation was analyzed to determine its critical path. The simulation setup with the Device Under Test (DUT) is depicted in Fig. 4.7. The configuration of the connection between the FAs in the chain was selected to strain the chain considerably. A 5k point Monte-Carlo transient simulation was performed to measure delay, as shown in Fig. 4.8. As expected, the low-voltage dynamic FA failed on most of the tests. When it did succeed, its performance was poor (with a higher mean delay than CMOS) and it displayed the highest overall variance. On the other hand, dynamic DML exhibited the lowest overall delay. The CMOS and static mode DML presented very similar slower delays.

To analyze performance and energy consumption, we constructed a 20-bit ripple adder using the FAs illustrated in Fig. 4.7. The delay–energy (E–D) analysis of a single block of the ripple adder chain is presented in Fig. 4.9. As expected, static DML consumed less energy than CMOS with degraded performance, while

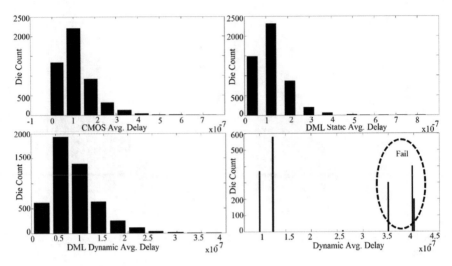

Fig. 4.8 Monte-Carlo FA delay analysis, $V_{DD} = 0.3\,\text{V}$. $\mu_{CMOS} = 155\text{n}$, $\sigma_{CMOS} = 84\text{n}$. $\mu_{Dynamic} = 150\text{n}$, $\sigma_{Dynamic} = 141\text{n}$. $\mu_{DML\_Static} = 190\text{n}$, $\sigma_{DML\_Static} = 80\text{n}$. $\mu_{DML\_Dynamic} = 96\text{n}$, $\sigma_{DML\_Dynamic} = 48\text{n}$

dynamic DML consumed less energy than dynamic logic but achieved boosted performance as compared to its CMOS counterpart. Furthermore, in all cases the MEPs were located in the ST region [24, 25]. Interestingly, the operation at the MEP of the static DML (250 mV) had an iso-frequency of 1.62 MHz with less than 0.72 fJ of energy consumption, whereas switching to the dynamic mode provided roughly a 2.3× frequency boost to 3.69 MHz with less than a 40% increase in energy consumption. To verify the parameter estimation of the transregional model, we compared the analytical delay graphs versus the measurement delay graphs for different voltages. These showed no practically visible error for voltages exceeding 300 mV and an average least square fit error of less than 1–2% (for all 1–3 transistor stacks). This implies that the model was very accurate.

The activity factor for the E–D graph shown in Fig. 4.9 was set at 1. An activity factor typically reveals tradeoffs with respect to switching and leakage energy and serves to analyze them separately. This activity factor was selected to represent the worst-case scenario for DML gates (in terms of the objective of energy). At all times and in all operation modes, DML logic had at least one minimum-sized transistor network. The energy dissipation in these voltage regions is known to be dominated by the subthreshold and gate leakage currents which are linearly dependent on the transistor widths, thus making the DML leakage substantially lower. Therefore, an activity factor of 1 constitutes a stringent criterion and is a pessimistic expectation.

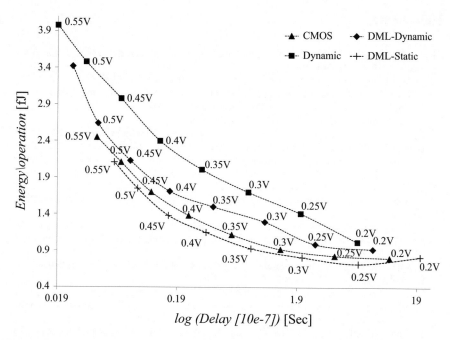

Fig. 4.9 FA E–D plots (per operation) of CMOS, dynamic, DML dynamic, and DML static as a function of V_{DD}

4.5   Conclusion

To provide the reader with a better understanding of DML circuits, this chapter discussed DML gate operation and optimization for low-voltage operations. It detailed device sizing analyses and optimizations for the ST and NT regions. It demonstrated simulated and fabricated devices, DML chains of NORs/NANDs, and FAs and compared them to their CMOS and domino counterparts. Finally, we reviewed the robustness characteristics of DML (through the SNM metric) from near- to subthreshold voltages. Monte-Carlo simulations were conducted and showed that in the dynamic mode, the DML FA was more robust and consumed less power than domino logic and was faster than the CMOS gate with a higher power value. Specifically, dynamic operation of the DML FA provided a 200% better performance than CMOS, whereas the static operation provides improved power consumption with reduced speed compared to its CMOS counterpart. Starting in the next chapter we extend this "gate-level" analysis to aspects of the system-level architecture and control mechanisms for DML.

References

1. A.P. Chandrakasan, S. Sheng, R.W. Brodersen, Low-power cmos digital design. IEICE Trans. Electron. **75**(4), 371–382 (1992)
2. C. Piguet, *Low-power Electronics Design* (CRC Press, Boca Raton, 2018)
3. P. Kalyani, P.S. Kumar, P.C. Sekhar, Design of subthreshold adiabatic logic based combinational and sequential circuits, in *2017 International Conference on Emerging Trends & Innovation in ICT (ICEI)* (IEEE, Piscataway, 2017), pp. 9–14
4. S.B. Nasir, S. Sen, A. Raychowdhury, Switched-mode-control based hybrid ldo for fine-grain power management of digital load circuits. IEEE J. Solid-State Circuits **53**(2), 569–581 (2017)
5. S. Kim, S.-Y. Lee, S. Park, K.R. Kim, S. Kang, A logic synthesis methodology for low-power ternary logic circuits. IEEE Trans. Circ. Syst. I Regul. Pap. **67**(9), 3138–3151 (2020)
6. Y. Chen, H. Jiao, Standard cell optimization for ultra-low-voltage digital circuits, in *2019 International Conference on IC Design and Technology (ICICDT)* (IEEE, Piscataway, 2019), pp. 1–4
7. J. Kwong, Y.K. Ramadass, N. Verma, A.P. Chandrakasan, A 65 nm sub-microcontroller with integrated sram and switched capacitor dc-dc converter. IEEE J. Solid-State Circuits **44**(1), 115–126 (2009)
8. D. Markovic, C.C. Wang, L.P. Alarcon, T.-T. Liu, J.M. Rabaey, Ultralow-power design in near-threshold region. Proc. IEEE **98**(2), 237–252 (2010)
9. G. Gammie, A. Wang, H. Mair, R. Lagerquist, M. Chau, P. Royannez, S. Gururajarao, U. Ko, Smartreflex power and performance management technologies for 90 nm, 65 nm, and 45 nm mobile application processors. Proc. IEEE **98**(2), 144–159 (2010)
10. B. Zhai, S. Hanson, D. Blaauw, D. Sylvester, Analysis and mitigation of variability in subthreshold design, in *Proceedings of the 2005 International Symposium on Low Power Electronics and Design* (ACM, New York, 2005), pp. 20–25
11. G. Gammie, A. Wang, M. Chau, S. Gururajarao, R. Pitts, F. Jumel, S. Engel, P. Royannez, R. Lagerquist, H. Mair et al., A 45nm 3.5 g baseband-and-multimedia application processor using adaptive body-bias and ultra-low-power techniques, in *2008 IEEE International Solid-State Circuits Conference-Digest of Technical Papers* (IEEE, Piscataway, 2008), pp. 258–611
12. B.H. Calhoun, A.P. Chandrakasan, Ultra-dynamic voltage scaling (udvs) using sub-threshold operation and local voltage dithering. IEEE J. Solid-State Circuits **41**(1), 238–245 (2006)
13. M.-E. Hwang, K. Roy, Abrm: Adaptive-ratio modulation for process-tolerant ultradynamic voltage scaling. IEEE Trans. Very Large Scale Integr. (VLSI) Syst. **18**(2), 281–290 (2010)
14. M.-A. LaCroix, H. Wong, Y.H. Liu, H. Ho, S. Lebedev, P. Krotnev, D.A. Nicolescu, D. Petrov, C. Carvalho, S. Alie et al., 6.2 a 60gb/s pam-4 adc-dsp transceiver in 7nm cmos with snr-based adaptive power scaling achieving 6.9 pj/b at 32db loss, in *2019 IEEE International Solid-State Circuits Conference-(ISSCC)* (IEEE, Piscataway, 2019), pp. 114–116
15. G. Papadimitriou, A. Chatzidimitriou, D. Gizopoulos, Adaptive voltage/frequency scaling and core allocation for balanced energy and performance on multicore cpus, in *2019 IEEE International Symposium on High Performance Computer Architecture (HPCA)* (IEEE, Piscataway, 2019), pp. 133–146
16. M. Pons, C.T. Müller, D. Ruffieux, J.-L. Nagel, S. Emery, A. Burg, S. Tanahashi, Y. Tanaka, A. Takeuchi, A 0.5 v 2.5 μw/mhz microcontroller with analog-assisted adaptive body bias pvt compensation with 3.13 nw/kb sram retention in 55nm deeply-depleted channel cmos, in *2019 IEEE Custom Integrated Circuits Conference (CICC)* (IEEE, Piscataway, 2019), pp. 1–4
17. H. Soeleman, K. Roy, B. Paul, Sub-domino logic: ultra-low power dynamic sub-threshold digital logic, in *Fourteenth International Conference on VLSI Design, 2001* (IEEE, Piscataway, 2001), pp. 211–214
18. I.E. Sutherland, R.F. Sproull, D.F. Harris, *Logical Effort: Designing Fast CMOS Circuits* (Morgan Kaufmann, Boston, 1999)

19. J. Keane, H. Eom, T.-H. Kim, S. Sapatnekar, C. Kim, Subthreshold logical effort: a systematic framework for optimal subthreshold device sizing, in *Proceedings of the 43rd Annual Design Automation Conference* (ACM, New York, 2006), pp. 425–428

20. D.M. Harris, B. Keller, J. Karl, S. Keller, A transregional model for near-threshold circuits with application to minimum-energy operation, in *2010 International Conference on Microelectronics* (IEEE, Piscataway, 2010), pp. 64–67

21. A. Kaizerman, S. Fisher, A. Fish, Subthreshold dual mode logic. IEEE Trans. Very Large Scale Integr. (VLSI) Syst. **21**(5), 979–983 (2013)

22. C.-H. Chang, J. Gu, M. Zhang, A review of 0.18-/spl mu/m full adder performances for tree structured arithmetic circuits. IEEE Trans. Very Large Scale Integr. (VLSI) Syst. **13**(6), 686–695 (2005)

23. J. Kwong, A.P. Chandrakasan, Variation-driven device sizing for minimum energy subthreshold circuits, in *Proceedings of the 2006 International Symposium on Low Power Electronics and Design* (ACM, New York, 2006), pp. 8–13

24. B.H. Calhoun, A. Wang, A. Chandrakasan, Modeling and sizing for minimum energy operation in subthreshold circuits. IEEE J. Solid-State Circuits **40**(9), 1778–1786 (2005)

25. S. Hanson, B. Zhai, K. Bernstein, D. Blaauw, A. Bryant, L. Chang, K.K. Das, W. Haensch, E.J. Nowak, D.M. Sylvester, Ultralow-voltage, minimum-energy cmos. IBM J. Res. Dev. **50**(4.5), 469–490 (2006)

Chapter 5
DML Energy-Delay Tradeoffs and Optimization

As shown in the previous chapters, DML design provides very high energy-delay (E-D) optimization flexibility at the gate level. In this chapter, this flexibility is utilized to enhance the energy efficiency and performance of larger combinatorial circuits. In other words, we go up the design hierarchy to the (small) block level. The goal is to overview DML energy-delay tradeoffs for a composite block and present solutions that capitalize on the DML's unique structure to achieve energy reduction and performance improvement. Specifically, we present critical-path-DML approaches that analyze the design's critical paths and selectively allow for their operation in the fast, dynamic mode; by contrast, the energy reduction is achieved by static operation of the non-critical path of the system. These approaches are demonstrated on a Carry Look-Ahead DML adder example. The analysis is carried out as a function of supply voltage and the operand size of the adder (n).

5.1 Introduction: Static DML as a Semi-Energy-Optimal CMOS

The design space of a CMOS gate is primarily influenced by V_T, the transistor width, V_{DD}, the channel length, oxide thickness, and body voltage. The impact of these parameters on E-D plane optimization has attracted considerable attention. In the CMOS family, the symmetry of the gate (i.e., equal rise and fall times) is crucial since in a combinational system there is always some uncertainty as to the transition, which is data-dependent, delay-dependent, parasitic, and implementation-dependent. Thus, the pull-up network (PUN) of CMOS gates, which is constructed with low mobility PMOS devices,[1] is sized up by the β parameter [1]. When

---

[1] In many modern nano-scale technologies, such as FinFET, the strength of the PMOS devices can be very similar to the strength of the NMOS devices.

© Springer Nature Switzerland AG 2021
I. Levi, A. Fish, *Dual Mode Logic*, https://doi.org/10.1007/978-3-030-40786-5_5

optimizing CMOS gate energy to the detriment of its performance, the transistor width is the prime parameter for reducing energy consumption, for a number of reasons that are enumerated below:

1. The switching energy is proportional to the load and quadratically dependent on V_{DD}. Under energy optimization, symmetry in gate performance does not constitute a constraint so that the transistor width can also be made smaller as well as β. This lessens the load capacitances considerably.
2. With the advent of V_{DD} lowering and technology scaling, leakage energy has become one of the key sources of total power dissipation. Leakage energy is caused by numerous leakage currents in the device but the main ones are the sub-threshold and gate leakage currents [2]. Thus, in the case of energy optimization, the transistor width can be reduced significantly, as well as the β of the gate.

CMOS-based DML gates operated in the static mode with transistor sizes opti-mized for the dynamic mode are *de-facto* a semi-energy-optimal CMOS structure with an additional negligible output capacitance for clocked transistors (transistors $M1$ and $M2$ in Fig. 2.1). In fact, when optimizing sizes for the dynamic mode, the complementary network is composed of minimum sized gates so that in the static mode these minimum sized transistors have minimal dynamic and static energy consumption (currents through the gate). Static DML is still highly robust because of its complementarity [3, 4] and can withstand aggressive voltage scaling, as discussed above. Thus, this methodology can also be seen as a stand-alone semi-optimal technique for reducing the energy consumption of digital circuits in general.

5.2 Critical-Path-DML Approaches to Energy Efficiency and High Performance

This section describes approaches to the energy efficient and high performance design of combinatorial systems. We start by presenting an approach that utilizes DML gates in the dynamic mode on the CPs to improve their delays. Then, we turn to factors affecting energy reduction in all the non-CP portions of the design.

Theoretically, a general DML design can be controlled (data-driven or external-signal-driven control) to operate each gate in either the static or the dynamic mode. What this means is that a general design can be operated in 2^g different ways, where g denotes the number of gates in the design. Each mode leads to a different operating point on the E-D space of the design. Figure 5.1a visualizes this modularity. The degenerated approaches for operating all the gates in one of the two modes, are shown in Figs. 5.1b and c. Switching between modes results in a clear-cut tradeoff since the design is optimized to either achieve maximum performance or minimum energy consumption.

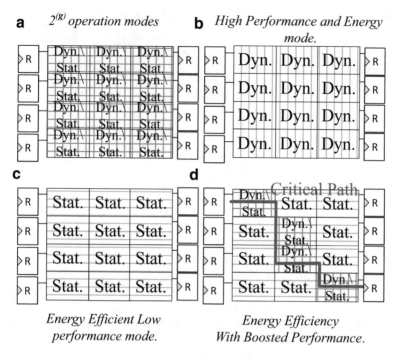

Fig. 5.1 DML-based design modes: (**a**) DML design optional operation modes, (**b**) DML design degenerated to the dynamic mode, (**c**) DML design degenerated to the static mode, (**d**) DML design where only the CPs are dynamically operated, while the rest of the design operates in the low energy static mode, where Dyn. stands for Dynamic and Static stands for Static DML or CMOS logic

5.3 Solution for Critical Path Timing Violations and Energy Consumption Reduction

Standard design flow tools automatically identify the CPs of a design. By replacing these paths with DML gates and applying the DML dynamic mode in these paths, their delay can be shortened. The rest of the design can be implemented using standard CMOS static logic. Clearly, special design constraints need to be enforced at all the intersections between the static paths and dynamic paths. In some of these cases, a footer should be applied [3–6]. Figure 5.1d illustrates a design in which the CPs were located and only those paths were allowed to toggle between the dynamic and static modes as a function of the system requirements. Whereas under some workloads the system can withstand slower operation, the CP logic will operate in the static mode. By contrast, when for other workloads the system must meet a clock with a shorter period, the CPs will operate in the dynamic mode. Normally, the number of gates on the CP is low as compared to the total number of gates in the design. Thus, in most cases, the inherent dynamic-operation energy of these CPs will not lead to a significant increase in the total energy consumption of the design.

Fig. 5.2 A mapped circuit
for CPs and non-CPs when
the former is operating in the
dynamic mode and the latter
in the static mode

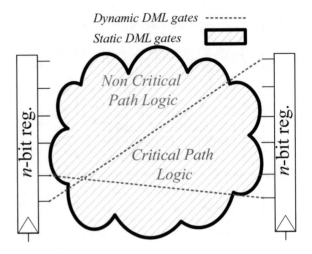

As noted in the previous paragraph, the CPs are mapped and operated in the
dynamic DML mode. In the paragraph above, the rest of the circuit was assumed to
preserve a standard CMOS logic gate topology. This explains why the design was
built to solve the CPs' timing constraints at the expense of a slight degradation in
energy consumption, as compared to a complete CMOS design. Next, we describe a
different approach in which all the components of the design that are not part of the
CPs are mapped to static mode DML gates (similar to the semi-energy optimized
CMOS gates described in the previous section). In most designs, these non-CPs
are not time-constrained so that the asymmetric behavior of their transitions and by
extension their performance degradation do not affect the clock period. The use of
the static DML mode for the vast majority of gates in the design leads to a notable
reduction in the total dynamic and static energy consumption. Figure 5.2 charts this
approach.

5.4   Modular Benchmark Example: Carry Save Adder
Design

In this section, we detail the benchmark chosen to illustrate the approaches
discussed in this chapter so far. The design can be operated in one of the three
modes:

1. A CP acceleration, which has two operation options:

   - "DML Carry Path-Dynamic"—The DML CPs are activated in the dynamic
     mode.
   - "DML Carry Path-Static"—The DML CPs are activated in the static mode.

   In both, the rest of the non-CP portions of the system are designed with standard
   CMOS.

2. A CP acceleration mode with low energy consuming non-CPs, which has two operation options:

   • "DML Carry Path-Dynamic with low energy non-CPs-Static"—The DML CPs are activated in the dynamic mode, while the rest of the system operates in the DML static mode.
   • "DML Carry Path-Static with low energy non-CPs-Static"—The DML CPs are activated in the DML static mode, similarly to the rest of the system.

3. CMOS equivalent design.

A Carry Save Adder (CSA; also called a carry bypass adder in some works) was chosen as the benchmark to demonstrate and evaluate this concept. The CP of the CSA increases in length as a function of the number of inputs, making it a simple candidate to evaluate E-D trends as a function of the CPs' lengths. Crucially, note that these methods can be applied to any combinatorial circuit and that CSA was chosen as a benchmark solely for its modularity and simplicity.

5.4.1 The CMOS CSA Design

A conventional CSA is composed of a set of Ripple Carry Adder (RCA) blocks. They utilize the carry propagation to skip the carry from one RCA block to the next RCA block. Because the propagation of the carry by a simple XOR gate can be predicted [7], the delay can be substantially reduced [8]. The CP in CSA is active when the carry ripples through the first block, then skips the rest of the middle blocks, and then ripples again through the last block. This is the longest possible route in the CSA. Lehman *et al.* investigated CSAs that had a non-uniform size (the same input size) for the RCA blocks [9]. Majerski presented a multi-level carry-skip propagation architecture [10]. Guyot *et al.* and Oklobdzija *et al.* put forward algorithms for choosing optimized block sizes [11, 12]. Other enhancements and implementations of advanced CSAs have been investigated in recent publications, e.g., [13–15]. For simplicity's sake, a CMOS CSA design with a fixed block size of 4 bits was implemented, as shown in Fig. 5.3. Again, the methods presented in this section can be generalized to any CSA block size constant or variable and any multi- or single-level carry path. The general single-bit FA equations are

$$S = A \oplus B \oplus C_{in}, \tag{5.1}$$

$$C_{out} = AB + C_{in}(A + B), \tag{5.2}$$

$$P = A \oplus B, \tag{5.3}$$

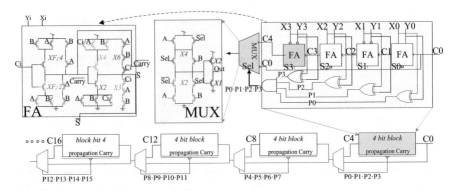

Fig. 5.3 Test-bench circuit: CMOS designed CSA where the non-elementary CMOS gates such as the mirror FA and MUX are presented with transistor sizes. Transistor sizes are marked by an X'Number', where X represents the multiplication by W_{min} (minimum transistor width). P and C stand for the propagate and carry signals

where $\oplus$ is the conventional XOR symbol. For an RCA, C_{out} will be an input to the next FA. For the CP, the carry will propagate through all FAs. Given the fact that C_{out} is on the CP for each RCA, the mirror circuit for computing C_{out} is used [8], as shown in Fig. 5.3. This circuit calculates the inverted value C_{out}, and when serially chained, it reduces the circuitry on the CP (i.e., it eliminates one inverter for each FA). Furthermore, the use of the mirror adders makes it obligatory to have inverting inputs for all odd FAs and inverting outputs for all even FAs [7], as shown in Fig. 5.3. All the logical gates presented in the figure are constructed with standard CMOS. A standard sizing optimization for the RCA of mirror FAs using Logical Effort [16] yields the sizing factor F_i (as shown in Fig. 5.3 for all the carry path gates) for all is, which are a multiple of 4, $F_i = 1$ and for all the rest $F_i = 3.5$.

5.4.2   DML Critical Path Design

Figure 5.4 shows the DML implementation of the CSA's CP. The CP flows through the first NOR (under the assumption that the carry in of the whole design is "0") and through all the MUXs of the design. The gate-level implementation of the CP can be constructed with numerous topologies of DML, but it is worth noting that DML NOR gates are more efficiently implemented in *Type-A* topologies and NAND gates in *Type-B*, as discussed in the previous chapters (and in [3–5]). Boolean logic does not allow for an efficient implementation of a MUX with a NOR following a NAND or vice versa, which is the preferred topology for DML logic design. Thus, in this topology, the CP is composed solely of NANDs (where one is implemented using an efficient *Type-B* and the other has a less optimal *Type-A* structure). The last inverter in each RCA block is a headed *Type-B* inverter that maintains the correct precharge

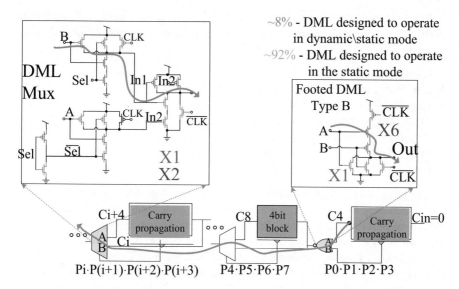

Fig. 5.4 DML critical path design for the benchmark CSA. The figure shows transistor widths for the gates of interest

phase for the CP. The sizes of the transistors in terms of minimal transistor width are shown in Fig. 5.4. In the design, when implemented in this way, only 8% of the transistors will (optionally) operate dynamically, and the remaining 92% stay in the low energy static mode. This modular design has the same complexity and the same dynamic-to-static-gates ratio for different input vector lengths of N bits.

5.5 Energy-Delay Plane as a Function of V_{DD} and n

The modular benchmark circuits described in the previous section were simulated in a standard 40nm CMOS process using a Cadence Virtuoso simulator. The modes (and methods) of the benchmark CSAs are primarily illustrated over the E-D plane and as a function of the CP length and the operating voltage. Note that the naming conventions for the different designs and operating modes are listed in Sect. 5.4. All the energy and delay metrics provided below are per-operation.

5.5.1 The E-D Plane as $f(V_{DD})$

To examine the approaches discussed above for low voltage and strong inversion operations, evaluations were carried out with supply voltages ranging from 0.4V to 1.1V.

The E-D curves for all designs of a 128-bit CSA are plotted in Fig. 5.5. As shown in the figures, the order of the curves from top to bottom represents the following designs:

1. CMOS.
2. CMOS design with a CP in the dynamic DML mode.
3. CMOS design with a CP in the static DML mode.
4. DML design with low energy static-DML mode operation for the non-CPs and CPs operating in the dynamic DML mode.
5. DML design with low energy static-DML mode operation for the non-CPs and CPs operating in the static DML mode.

The latter two curves are in the lower region of the plane which represents the low energy achieved by implementing all non-CPs in the low energy DML static mode (which, as described above, can also be referred to as *a low energy CMOS flavor*). The two areas of interest are circled on the edges of Fig. 5.5a and are enlarged in Fig. 5.5b and c. Figure 5.5b shows the tradeoff area for a 1.1V operating voltage for all designs. Figure 5.5c presents the same tradeoff for 400mV power supply. These two extremes clearly show that these designs are highly flexible in terms of energy consumption and performance for the whole range of voltages. This comparison of the DML enhanced CP plots (second and third curves) to the CMOS plot (first curve) for the 0.4V supply (Fig. 5.5c) clearly shows that the DML enhanced CP exhibits a significant improvement in performance. This achievement comes nevertheless at the cost of a 16% increase in energy consumption. For example, consider a multi-frequency system where the module's frequency can vary in time. When a low power operation is required, the static mode (with a low frequency) can be applied, yielding 2.5X energy improvement with a penalty of a performance drop of 1.3X. This ability to change operating conditions on the E-D plane on-the-fly is a feature that can easily be exploited to improve system flexibility and E-D efficiency.

For the 1.1V supply (Fig. 5.5b), boosting the performance of the CP by 20% only increases energy consumption by 3%. When a low power operation is needed, the static mode can be applied, which leads to a 1.5X energy improvement at the cost of a performance drop of 1.4X. These results help show that a low voltage operation magnifies the differences between the different modes which can be accounted for as follows. The first is that the performance superiority of DML circuits in the dynamic mode over standard CMOS intensifies with a drop in supply voltage lowering [3–5]. The second, less dominant factor, is the reduced sensitivity of DML circuits to increased leakage currents at low supply voltages [3–5].

Inspecting the DML performance optimized CP with low energy non-CP plots (the two lowermost curves), makes it clear that the total energy drops two- to three-fold for all voltage regions. Moreover, the 1.3X and 2.1X improvement in CP performance was also found for the 1.1V and 400mV supplies, respectively. These CP performance results are comparable to the results obtained for operation without the low energy non-CP gates. This is because the CPs themselves have not changed. Thus, overall, the flexibility of the DML design leads to significant improvements in both energy and performance.

Fig. 5.5 E-D measurements for all 128-bit designs. (**a**) $E - log(D)$ plots for all supply voltages from 0.4V to 1.1V. (**b**) Enlarged E-D plots for 0.4–0.5V supply voltages. (**c**) Enlarged E-D plots for 1–1.1V supply voltages

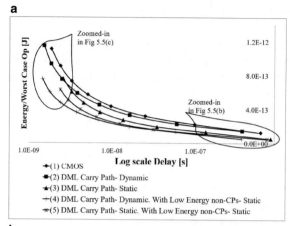

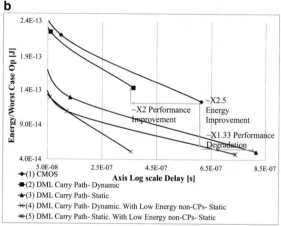

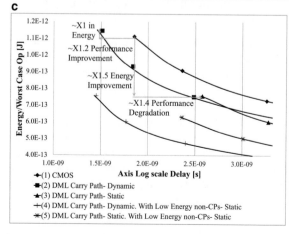

5.5.2   The E-D Plane as $f(N)$

This section examines the efficiency of the concept as a function of the CP's length, which is closely related to the size of the design. The CSA's size\length depends on the number of input bits, N. Figure 5.6 depicts the E-D trends for all designs as a function of N. Each plot starts with the minimal CP associated with a size of $N = 4$ and goes up to the longest examined CP of $N = 128$; namely, the point where $N = 128$ appears in both Figs. 5.5 and 5.6. The key purpose of this analysis is to show the scalability of the method for various design sizes and not only for a very long CP. Figure 5.6a and b demonstrates that as N increases, the ratio between the energy/performance of the different designs is almost constant. Thus, the design remains fully scalable for 400mV and 1.1V. Figure 5.6b also depicts another interesting feature of the 128-bit design with $V_{DD} = 1.1V$: the low energy design (DML static mode for non-CPs) with CPs operated in the dynamic mode consumes slightly more energy than the standard CMOS non-CP design with DML dynamic CP but achieves more than a 2X improvement in performance.

As shown in Fig. 5.6a, all designs ($N = 4...128$) with performance-enhanced CPs showed a significant improvement in performance at 400mV compared to their CMOS counterparts. However, for the 1.1V supply (Fig. 5.6b), this efficiency was only observed from $N = 32$. This behavior naturally depends on the specific gate topology of the chain, as discussed in the previous chapter. The specific CSA design represents an average case where some of the DML gates on the CP are very fast compared to CMOS, such as *Type-B* NAND, and others make very small improvements, such as *Type-A* NAND. For this reason, we expect that for other benchmarks, the improvement in E-D will be clear-cut for some $N > N_{MIN}$.

5.5.3   Stimuli Input Vector Complexity

The results presented in the previous subsections examined input stimuli that activated the CP of each circuit. These stimuli triggered the worst delays possible for these designs since each circuit requires different inputs to activate its CP. The worst case scenario for energy consumption also occurs with a specific input vector when it switches as many gates as possible for each RCA chain (the static portions of the design). In the previous two subsections, for the case of 128-bit CSA, the input vectors were chosen to switch 40 outputs independently of the CP switching. This arrangement is highly unlikely, since the average number of switching outputs is generally below 40. Let us assume equal probabilities for a logic "1" and a logic "0" for each input. The probability of a carry in a FA is $q = 0.5$. The probability for a carry to propagate through K successive bits is

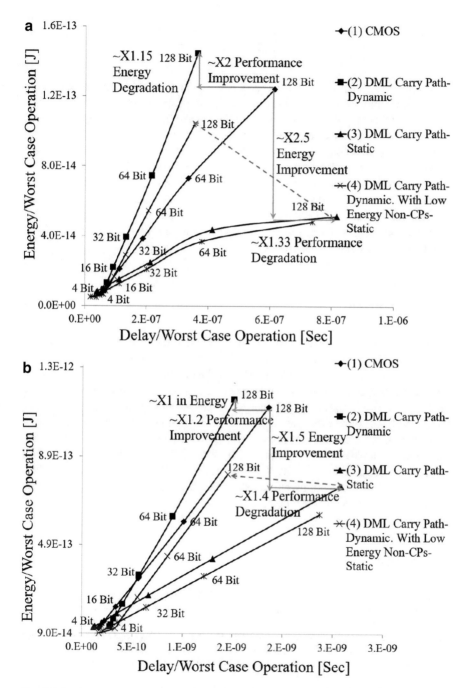

Fig. 5.6 E-D measurements as a function of the CSA size. (**a**) For 0.4V. (**b**) For 1.1V.

$$q_k = \underbrace{\frac{1}{2} \cdot \frac{1}{2} \cdot \frac{1}{2} \cdots \frac{1}{2}}_{K} = \frac{1}{2^k}. \qquad (5.4)$$

Alternatively, the probability of a carry being either "killed" or generated through K successive bits is $1 - q_k$. Thus, the probability of propagating more than 4 bits is 6.25%, which is quite low. For example, let us take the 128-bit design composed of 4 bit RCAs (i.e., 32 segments): the rippling of 2 bits within each 4 bit RCA (in addition to the switching of the whole CP) is a quite reasonable or even a stringent case in terms of probability. As anticipated when input vectors that are more energy-consuming for the static parts of the design were simulated (60 and 80 switched outputs), the input stimulus complexity rose and the additional energy required for the dynamic operated CP became increasingly negligible in comparison to the total energy of the designs. These results are reassuring for all worst-, typical-, and best-case input vectors in terms of energy.

5.6   Conclusion

Timing closure and energy minimization are critical issues in all digital circuits. The vast possibilities inherent to designs with DML gates leverage the flexibility of the design to meet CP timing constraints while reducing the total energy consumed by the circuit, as demonstrated in this chapter. Today, the CP timing issue is closely related to the rise in the consumed energy associated with conventional methods. In this chapter, we counter this paradigm by showing that both timing and low energy consumption requirements can be met. We showed that the performance of the 40-nm CSA benchmark circuit is improved by 2X, while also achieving a reduction in energy consumption of 2.5X. Since the CSA circuit is not optimal for DML implementations, these improvements should be even greater for other designs, as is shown in the following chapters.

References

1. M.K. Stojčev, J.M. Rabaey, A. Chandrakasan, B. Nikolić, *Digital Integrated Circuits: A Design Perspective*, 2nd edn. Facta Universitatis-Series: Electronics and Energetics, vol. 16(1), pp. 155–157, 2003
2. S.P. Mohanty, N. Ranganathan, E. Kougianos, P. Patra, *Low-Power High-Level Synthesis for Nanoscale CMOS Circuits* (Springer Science & Business Media, 2008)
3. A. Kaizerman, S. Fisher, A. Fish, Subthreshold dual mode logic. IEEE Trans. Very Large Scale Integr. VLSI Syst. **21**(5), 979–983 (2013)
4. I. Levi, A. Kaizerman, A. Fish, Low voltage dual mode logic: Model analysis and parameter extraction. Microelectronics J. **44**(6), 553–560 (2013)

5. I. Levi, A. Belenky, A. Fish, Logical effort for cmos-based dual mode logic gates. IEEE Trans. Very Large Scale Integr. VLSI Syst. **22**(5), 1042–1053 (2014)
6. I. Levi, O. Bass, A. Kaizerman, A. Belenky, A. Fish, High speed dual mode logic carry look ahead adder, in *2012 IEEE International Symposium on Circuits and Systems* (IEEE, 2012), pp. 3037–3040
7. I. Koren, *Computer Arithmetic Algorithms* (Universities Press, 2002)
8. A.T. Tran, B.M. Baas, Design of an energy-efficient 32-bit adder operating at subthreshold voltages in 45-nm cmos, in *2010 Third International Conference on Communications and Electronics (ICCE)* (IEEE, 2010), pp. 87–91
9. M. Lehman, N. Burla, Skip techniques for high-speed carry-propagation in binary arithmetic units. IRE Trans. Electron. Comput. **EC-10**(4), 691–698 (1961)
10. S. Majerski, On determination of optimal distributions of carry skips in adders. IEEE Trans. Electron. Comput. **EC-16**(1), 45–58 (1967)
11. A. Guyot, B. Hochet, J.-M. Muller, A way to build efficient carry-skip adders. IEEE Trans. Comput. **36**(10), 1144–1152 (1987)
12. V.G. Oklobdzija, E.R. Barnes, Some optimal schemes for alu implementation in VLSI technology, in *1985 IEEE 7th Symposium on Computer Arithmetic (ARITH)* (IEEE, 1985), pp. 2–8
13. S. Patel, B. Garg, A. Mahajan, S. Rai, Area-delay efficient and low-power carry skip adder for high performance computing systems, in *2019 IEEE International Symposium on Smart Electronic Systems (iSES) (Formerly iNiS)* (IEEE, 2019), pp. 300–303
14. R. Abinaya, S. Gayathri, S. Atchaya, G.H. Kumar, G.N. Balaji, Power efficient carry skip adder based on static 125nm cmos technology. Int. J. Innov. Res. Sci. Technol. **5**(8), 32–36 (2019)
15. B. Sanjana, K. Ragini, Design of a novel high-speed-and energy-efficient 32-bit carry-skip adder, in *Innovations in Electronics and Communication Engineering* (Springer, 2019), pp. 335–343
16. I.E. Sutherland, R.F. Sproull, D.F. Harris, *Logical Effort: Designing Fast CMOS Circuits* (Morgan Kaufmann, 1999)

Chapter 6
DML Control

This chapter focuses on the granularity of the DML mode control. First, we describe a coarse-grain data-dependent controller that controls DML at the block level. We show that the operation mode of DML can be selected by critical path prediction architectures that considerably enhance performance. Then, we present a design example of a fine-grain, data-dependent controller that operates at the logic path level. The main goal of this chapter is to show that DML primitives can be utilized to make improvements at numerous abstraction levels even though these are more often associated with the gate level. We provide several examples illustrating how primitives can be controlled as a function of gate, path, block, and architecture level requirements.

6.1 Coarse-Grain DML Mode Selection Controller

This subsection introduces the features related to the granularity of DML mode control signals. A sophisticated coarse-grain block-level data-dependent controller is discussed. We show that the DML operation mode can be selected by critical path prediction logic and treated at the architectural level, thus enhancing performance. For purposes of illustration we examine the dual mode square (DM^2) adder architecture as an example.

Achieving energy efficiency and low peak power while maintaining computational performance is one of the most sought-after goals of processor designs today. Energy reduction and performance improvement have been studied extensively starting from the very high level of application algorithms, through system [1], architecture [2, 3], and logic levels, to the gate [3–8], circuit, device, and interconnect levels [9, 10]. Energy reduction in the context of pipelined digital systems has also been investigated [3, 11]. For example, approaches such as circuit sizing and supply voltage scaling have been utilized and analyzed [3]. To combat energy cost, more recently, researchers are focusing on environmental-aware

© Springer Nature Switzerland AG 2021
I. Levi, A. Fish, *Dual Mode Logic*, https://doi.org/10.1007/978-3-030-40786-5_6

voltage and frequency scaling [12–14], on efficient power supply regulation and architectures [15–17].

This chapter combines two proposed gate and architecture level approaches. It shows how employing two separate methods leads to considerable performance enhancement and energy efficiency. The first method is known as dual-mode addition (DMADD) [18]. It takes advantage of the carry probability to perform low-power addition and results in a substantial energy reduction of up to 50% over conventional designs. However, it requires some pipeline modifications to support multi-cycle addition. The second method is the DML gate family and will draw on what we have learned from the previous chapters [19–23].

The chapter introduces the dual mode square (DM^2) approach that combines DMADD and DML. The main objective of DM^2 is to eliminate the need for multi-cycle addition in the DMADD by replacing its standard CMOS logic with DML, thus avoiding architectural overheads (multi-cycle). The main idea is to switch at the block level to the DML fast dynamic mode to complete the computation in a single cycle. DM^2 enables considerable energy savings related to the inherent properties of DML gates in the static mode (which operate most of the time). Two adders were implemented using the DM^2 method in a standard 40 nm process; as described below, the theoretical analysis and post-layout simulations showed that the DM^2 resulted in energy savings of up to 36%, as compared to the DMADD.

The DM^2 adder achieves low energy, high performance, and small area by combining these two independent techniques. The main ingredients are (1) on-the-fly adaptation of the gates to real-time system requirements by tapping DML's static and dynamic modes and their energy-performance tradeoff and (2) sophisticated control mechanisms to control these gates (using the probabilistic DMADD architecture).

The remainder of this subsection is organized as follows. We start by briefly presenting the DMADD and DML techniques. This is followed by the integration of the DMADD and DML into the DM^2 architecture, which incorporates a theoretical analysis and circuit design optimization. Finally, we support this architecture by describing simulations of 40 nm DM^2 adders and comparing them to standard CMOS-based DMADD, Brent–Kung, and Ripple adders.

6.1.1 Dual-Mode Addition (DMADD) Approach Overview

DMADD is made up of two addition modes [18]. The energy-efficient one-cycle mode, called *normal*, is used most of the time to properly compute addition. It takes advantage of the average (expected) longest carry-in addition, which is $O(log_2 n)$, which therefore is much shorter than the adder size n. The probability of an $O(n)$-bit carry propagation is nearly zero [24]. The second mode, called *extended*, occurs very infrequently and requires several clock cycles to properly add. The decision as to which mode should take place requires an appropriate control circuit. When this control is used in a pipelined processor, it selects the right mode at the instruction decode (ID) stage, prior to the arithmetic logic unit (ALU) stage.

The probability q of a carry to propagate through a bit is 1/2; thus, the propagation probability through successive k-bits is 2^{-k}. The probability q_k that it will take exactly k bits for a carry to either be generated or "killed" is:

$$q_k = \Pr\left(\prod_{i=1}^{k-1} p_i = 1\right) \times \Pr(p_k = 0) = 2^{-k}, \tag{6.1}$$

where p_i is the propagate signal of bit i. It was shown in [18] that adders designed for $2log_2n$-bit carry propagation have considerably more energy efficiency compared to ordinary n-bit carry propagation designs. An n-bit DMADD comprises groups of k bits each, where $n = mk$, such that the carry propagation delay of two k-bit adders meets the clock cycle. It makes several design alternatives to reduce energy dissipation. A design for a $(2k - 1)$-bit delay rather than n-bit enables transistor downsizing, high threshold voltage usage, or voltage scaling [1]. To compensate for cases where the carry propagates through more than $(2k - 1)$ bits, m clock cycles are used to complete the computation. The *normal* operation mode of the DMADD requires each m group to "kill" or generate a carry, for which the probability q_{norm} is:

$$q_{norm}(k, m) = \left(1 - 2^{-k}\right)^m = 1 - m2^{-k} + O\left(2^{-2k}\right) > 1 - m2^{-k}, \tag{6.2}$$

whereas the probability q_{ext} of the *extended* mode is:

$$q_{ext}(k, m) < m2^{-k}. \tag{6.3}$$

The deployment of DMADD in an in-order pipelined processor calls for stalling the pipe for m cycles in the case of extended-mode addition. This causes some design overhead and performance degradation. More critically, DMADD in out-of-order [25] architectures may be extremely difficult to implement. Here, we show how the utilization of DML avoids the *extended* multi-cycle mode by ensuring that regardless of the carry propagation, the DMADD will always compute properly within a single cycle.

6.1.2 Dual-Mode² (DM²) System Architecture and Transistor Sizing

6.1.2.1 DM² Architecture

The DM² adder example is an n-bit RCA divided into $m = n/k$ groups of k-bits each, as illustrated in Fig. 6.1. As discussed in the previous subsection, the

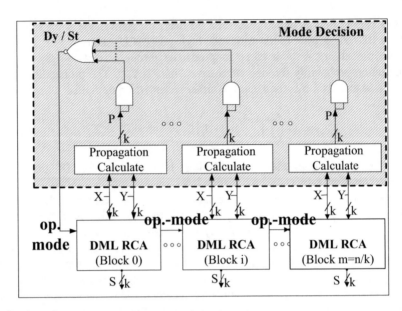

Fig. 6.1 DMADD adder topology and control circuit

probability of the *normal* addition mode is approximately $1 - m/2^k$. The longest carry path in this mode does not exceed $(2k - 1)$ bits, which is far shorter than the n-bit worst case. The underlying DML gates can therefore be operated in their static, energy-efficient mode. In its *extended* mode, where the carry propagates through more than $2k - 1$-bits with probability $m/2^k$, the DML logic will toggle to its fast dynamic mode. In this mode, the worst-case n-bit carry path must be completed within the given clock cycle, which is done by transistor sizing.

The tradeoff here is clear: most of the time the DMADD consumes very low power, and high power consumption is rare. Obviously, k must be defined such that the propagation delay of a $(2k - 1)$-bit carry path, where the logic is static, will not exceed the clock cycle. To minimize the dynamic mode probability, k is maximized as a function of this delay constraint.

Current-day processors are pipelined. To capture the advantages of DM^2, we used a simple, yet realistic in-order pipelined processor [25]. We took advantage of the fact that the ID stage occurs one cycle prior to execution, thus making the ALU arguments available one cycle ahead of their use. This also makes it possible to determine the operation mode of the DMADD by using a *mode decision block*, as illustrated in Fig. 6.2.

The mode decision block architecture is depicted in Fig. 6.1, where St and Dy denote the static-normal and dynamic-extended modes, respectively. The RCA is standard and is composed of alternating polarity FAs [26]. The alternating polarity of the RCA bits and the inherent DML alternating precharge polarity [19, 20, 23] dictate the differences in internal designs of the polarity alternating FA. Figure 6.3

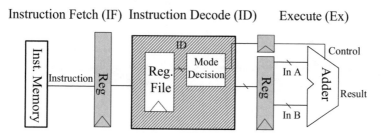

Fig. 6.2 Incorporation of mode decision logic

details the internal circuits of these two bit types. To speed up the critical path, it uses un-footed gates in even and odd bits.

It is crucial to note that the precharge of all the bits occurs simultaneously and hence does not affect the critical path delay. To ensure proper precharge, the gates connected to the RCA inputs are footed *Type-A* in the even bits and footed *Type-B* in the odd ones. The transistor-level schematics and the sizes of the alternating bits are shown in Fig. 6.3b. They are based on CCMOS (mirror) FA [26].

6.1.2.2 DM^2 Transistor Sizing

Let T be the system's clock cycle and T_{pre} the precharge delay of a full adder. The size of the DML gates is determined so that the carry evaluation through all the n bits will meet $T - T_{pre} = nT_{eval}$, where T_{eval} is the carry evaluation delay of a FA. Note that the precharge takes place simultaneously for all bits prior to evaluation, as illustrated in Fig. 6.4.

It is equally important to note that in DML, only the transistors involved in the evaluation network are susceptible to upsizing, whereas the other half of the complementary transistors stays minimal [19–23]. Furthermore, not all the evaluation transistors require upsizing, i.e., just those designated by S, as shown in Fig. 6.5, for the carry logic of two successive bits. The DML design methodology requires these to be opposite types.

As shown in Fig. 6.5a, the critical carry path in the *Type-A* gate passes through the lower left branch, whereas in the *Type-B* gate, it passes through the upper left branch. Consequently, the remaining evaluation transistors can remain minimal and are designated by 1. The smallest sizing factor S of the evaluation transistors that meet the timing constraints was based on a simulation.

Once the transistor sizes have been determined by the DML dynamic mode, the maximal group size complying with the timing constraints in the static mode can be set. Recall that in the DMADD normal mode, the static DML mode is operational, where the carry propagates through $2k - 1$-bits at most.

a
Gate level:

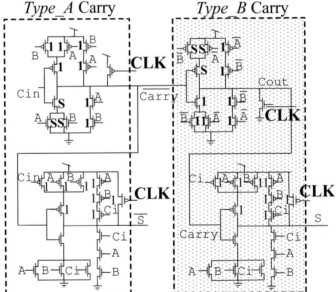

b
Transistor level:

Fig. 6.3 Even and odd FAs. At the (**a**) gate level (**b**) transistor level

The worst-case delay path for the static DM^2 mode is different from the dynamic path. Figure 6.5b illustrates the critical path in the static mode that passes through the highly resistive minimal size transistors.

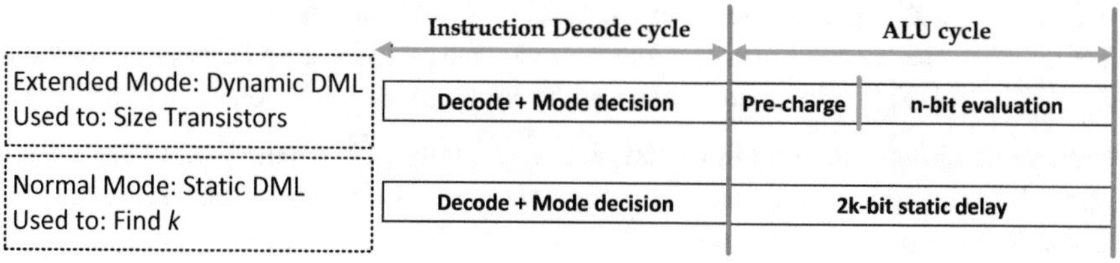

Instruction Decode cycle
ALU cycle
Extended Mode: Dynamic DML
Used to: Size Transistors
Decode + Mode decision
Pre-charge
n-bit evaluation
Normal Mode: Static DML
Used to: Find k
Decode + Mode decision
2k-bit static delay

Fig. 6.4 System timing diagram

a
Type_A Carry
Type_B Carry
b
Type_A Carry
Type_B Carry
Cin
CLK
VDD

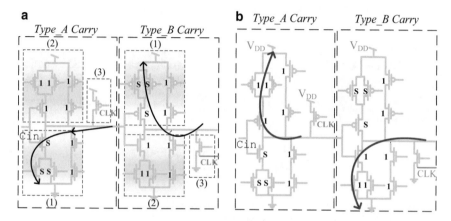

Fig. 6.5 DML worst-case delay path: (**a**) DML dynamic mode and (**b**) DML static mode, where the blocks represent (1) the evaluation path (2) the complementary networks, and (3) the precharge transistors

Let T_{stat} denote the carry delay in a FA operated in the DML static mode. The group size k is determined to satisfy $T = (2k - 1)\, T_{\text{stat}}$, yielding:

$$k = \frac{1}{2}\left(\frac{T}{T_{\text{stat}}} + 1\right) = \frac{1}{2}\left(\frac{T_{\text{pre}} + nT_{\text{eval}}}{T_{\text{stat}}} + 1\right). \tag{6.4}$$

Usually, n is a power of two, and for practical design considerations, k is set to the nearest power of two [18]. Because the size of the devices was chosen to be as small as possible, k must always be rounded down, since rounding up may cause timing violations.

6.1.3 Computing Energy Savings

Although the main rationale for DM^2 is to avoid the DMADD architecture overheads described above, it also leads to notable energy savings. To assess the DM^2 energy savings compared to DMADD, the latter was optimally designed in ordinary CMOS logic to meet the worst $2k - 1$-bit delay occurring in group size k. Once the size of the gates was determined, the switching and leakage energies, $E_{\text{switch}}^{\text{DMADD}}$ and $E_{\text{leakage}}^{\text{DMADD}}$, were measured by simulation. Let's consider DM^2 adder energy consumption per addition, and let $E_{\text{stat}}^{\text{DM}^2}$ and $E_{\text{dyn}}^{\text{DM}^2}$ be the worst-case static-normal mode (most often) and dynamic-extended mode (less often), respectively. Given that the normal mode probability is $1 - m/2^k$, we obtain:

$$\frac{E^{\text{DM}^2}}{E^{\text{DMADD}}} = \frac{\left(1 - m/2^k\right) E_{\text{stat}}^{\text{DM}^2} + m/2^k E_{\text{dyn}}^{\text{DM}^2}}{E_{\text{switch}}^{\text{DMADD}} + E_{\text{leakage}}^{\text{DMADD}} + (m-1)\, m/2^k E_{\text{leakage}}^{\text{DMADD}}}, \tag{6.5}$$

The term $(m-1)\, m/2^k E_{\text{leakage}}^{\text{DMADD}}$ in Eq. (6.5) follows from the extra $m-1$ cycles required by the DMADD extended mode. Equation (6.5) can be simplified by noting that $m/2^k \ll 1$, $E_{\text{dyn}}^{\text{DM}^2} \simeq 4 E_{\text{stat}}^{\text{DM}^2}$ for a DML FA (obtained by simulation), and that $E_{\text{switch}}^{\text{DMADD}} + E_{\text{leakage}}^{\text{DMADD}} \gg (m-1)\, m/2^k E_{\text{leakage}}^{\text{DMADD}}$. All in all, we obtain the following approximation:

$$\frac{E^{\text{DM}^2}}{E^{\text{DMADD}}} \simeq \frac{E_{\text{stat}}^{\text{DM}^2}}{E_{\text{switch}}^{\text{DMADD}} + E_{\text{leakage}}^{\text{DMADD}}} \tag{6.6}$$

Note that we did not include the energy consumed by the adder's controller since it is similar to DMADD and DM^2 adders.

6.1.4   Benchmark Results and Analysis

DMADD 32- and 64-bit adders were compared to a DM^2 adder. In addition, DM^2 performance was tested against ripple carry and Brent–Kung adders designed in a 40 nm process technology and targeting a 1 GHz clock frequency. Energy, area, and extended-mode probability and reliability were analyzed. As described in the previous subsection, the first DM^2 design step is to set the device sizes S of the FAs to meet the clock cycle in the DML dynamic mode, which defines the optimal group size k.

Note that the DM^2 adder requires extra circuitry for precharge, which should be carefully designed. In terms of the other components, the precharge circuits were carefully designed under PVT corners and mismatch, and the necessary margins were adhered to. The energy and delay overheads, which are negligible, are represented by the final results, as reported in this section.

6.1.4.1 Transistor Sizing and Setting the Group Size

Figure 6.6 presents the delays for the 32-, 64-, and 128-bit adders operating in both the dynamic and static modes. The delay of the dynamic mode decreases as S increases and is given by:

$$T_{\text{eval}} = (n - 1) \frac{R}{S} (\alpha C + \beta S C) + \frac{R}{S} (\delta C + \gamma S C), \qquad (6.7)$$

where R is the resistance and C is the capacitance of a minimal size transistor, and α, β and γ are process dependent parameters. The $(n - 1)$ factor represents the first $(n - 1)$ gate delays in the chain that charges similar capacitors, whereas the second term represents the last gate that charges the output register. Though not fully intuitive, the delay in the static DML mode increases with the increase in transistor size. This follows from the inherent structures of DML logic. Recall that in *Type-A* the size of the pull-up transistors through which the capacitive load is precharged is minimal. Similar arguments apply to the *Type-B* pull-down transistors (Fig. 6.5b). Overall, the static delay is given by:

$$T_{\text{stat}} = (2k - 1) R (\alpha C + \beta S C) + R (\delta C + \gamma S C). \qquad (6.8)$$

Consider the design of a 32-bit adder targeting a 1 GHz clock frequency. Ignoring setup time, the intersection point (a) of the dynamic curve with the 1 GHz horizontal line in Fig. 6.7 dictates the smallest sizing factor S that corresponds to the timing constraints. Ideally DM^2 should aim for the largest possible group size k, which results in the smallest probability of the dynamic (high energy) mode. This could

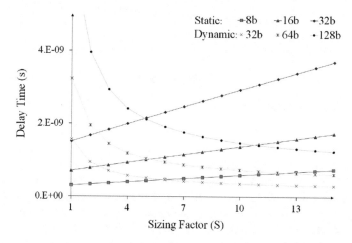

Fig. 6.6 Delays of 32-, 64-, and 128-bit DM^2 adders operated in both the dynamic and static modes

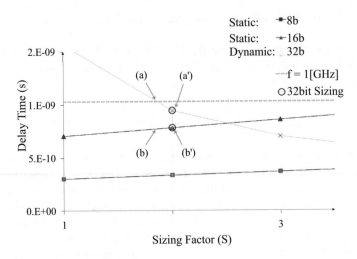

Fig. 6.7 32-bit adder design operation point

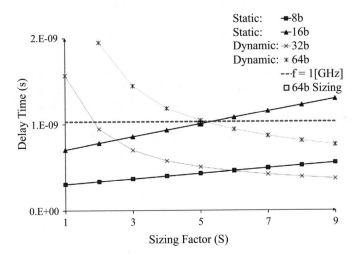

Fig. 6.8 64-bit adder design operation point

theoretically be achieved by the static curve passing through (a). Practically, since k is a power of two, it is obtained by the nearest match below point (a).

To mimic the common design methodology where sizing factors are integers and the DMADD group size is a power of two, the nearest practical design point (a') corresponding to $S = 2$ was chosen. The largest practical group size that met the delay constraints for $S = 2$ was (b'), yielding a 32-bit adder for which $2k = 16$.

Another 64-bit adder was designed with similar features, yielding $2k = 16$ and $S = 5$, as illustrated in Fig. 6.8 (the intersection of the dynamic, static, and clock cycle curves at a single point is purely a coincidence).

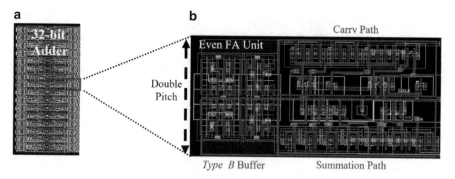

Fig. 6.9 (a) Layouts of a complete DM^2 adder occupying 400 μm^2 and a single FA cell in (b)

The procedure to find the minimal device sizing factor automatically determines the maximal group size, which in turn minimizes the dynamic DML operation mode probability. To summarize, the determination of the device sizes, on one hand, and the group size, on the other, is optimal via all means, and there is no other design point meeting the clock cycle that yields lower energy (Fig. 6.9).

The dynamic operation mode probability is derived by substituting n and k into Eq. (6.3), which for a 32-bit adder yields 1.56%, and 3.12% for a 64-bit adder.

6.1.4.2   Energy Saving Measurements and Bounds

To evaluate energy dissipation, the adder inputs were set such that the worst-case scenario of maximum energy consumption would occur. Figure 6.10 illustrates two successive bits of alternating types as defined by the DML design methodology. To trigger the worst case, the gates of all the evaluating devices in the *Type-A* cell should be at the "1" logic level. This is achieved by providing two "1" input bits and enforcing "1" in carry-in, which is obtained by providing two "0" logic levels to the inputs of a *Type-B* cell. The symmetric argument of ensuring that all the evaluating devices in the *Type-B* cell are conducting holds similarly. The gates of all the evaluating devices in the *Type-B* cell should be at the "0" logic level. This is obtained by providing two "0" input bits and enforcing "0" in the carry-in, by providing two "1" logic levels to the inputs of the *Type-A* cell.

Consequently, as shown in Fig. 6.3 and the alternating polarity of successive FA bits, the worst input of the adder is $\dfrac{A[0 : n - 1] \, 111111 \cdots 1111}{B[0 : n - 1] \, 111111 \cdots 1111}$. This worst-case input applies to both the static and dynamic DML modes. However, there is a considerable difference between them. Whereas the energy in the static mode is consumed by the evaluation devices alone, the dynamic mode consumes additional precharge energy, which was measured in the experiments. It is worth noting that the propagate signals of all the bits are "0"; therefore, since it is embedded in the pipeline, the DM^2 adder controller will turn it into a normal static mode. Bear in

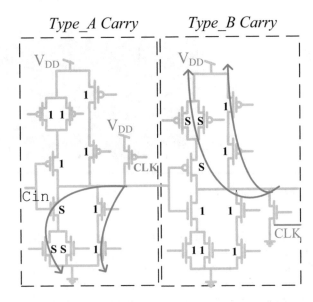

Fig. 6.10 Worst-case energy triggered paths in both the static-normal and dynamic-extended modes

mind that although we used this input to calculate the worst dynamic mode energy consumption, in the actual pipeline this scenario will operate in the static DML mode.

To compare the energy consumption of the ordinary DMADD CMOS adder to the DM^2, the worst-case stimuli were used for both adders. For the DMADD normal mode, the inputs resulted in the longest carry propagation through $2k - 1$ bits, whereas in the extended mode it propagated through the entire n bits. The worst stimulus for DM^2 is the one described previously.

Both DMADD and DM^2 adders were implemented in a 40 nm process technology. The layout of the 64-bit DM^2 adder is shown in Fig. 6.9. It was designed with Cadence Virtuoso tool and extracted and simulated with SPICE. The energy measure for each mode was weighted by its corresponding probability. The results are summarized in Fig. 6.11 and revealed a 36% energy reduction for the 32-bit DM^2 adder and a 27% reduction for the 64 bit compared to the corresponding DMADD adders.

Recall that the rationale for DM^2 design was to simplify the pipeline and avoid the multi-cycle mode required by the DMADD design. Note that there is no tradeoff in achieving the primary objective but that considerable energy savings are achieved.

6.1.4.3 Comparison of DM^2 to Brent–Kung and Simple Ripple Carry Adders

Extensive experiments were carried out on both the 32- and 64-bit adders to compare the average power per cycle (henceforth power) efficiency of DM^2 to a variety of

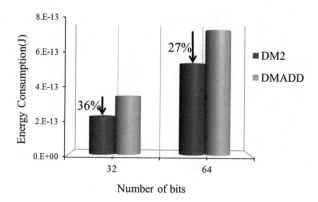

Fig. 6.11 Energy consumption for the 32-bit and the 64-bit adders

adder architectures. The experiments covered three architectures at the same target frequency: a high-performance Brent–Kung [27], a low-performance Ripple carry adder, and DMADD, which is the basic addition architecture [18] used for DM^2 (see the results reported above). As shown in Table 6.1, DM^2 achieved a power reduction of 1.9×–5×. All the adders were designed to meet 1 GHz performance, and their power consumption was minimized. Adders were Verilog designed and synthesized with the Register Transfer Level (RTL) compiler synthesis tool with the given 40 nm technology library and the Cadence encounter place and route capabilities. Then, all the designs were imported into Cadence Virtuoso for specter (SPICE) analog simulations. All the adders were simulated with their worst-case input transitions for power measurements and their slowest critical path frequency, as was implemented in the previous subsection for DM^2 and DMADD.

In the first experiment, the Brent–Kung architecture was used to meet the 1 GHz clock frequency; all the related attributes for the synthesis tool were set to minimize the power consumption. This resulted in a power consumption of ∼2.6× (593/230) and ∼2.1× (1082/520) for the 32- and 64-bit designs, respectively, compared to the DM^2 adder. Note that the Brent–Kung architecture is the only alternative if very high performance is required. For example, the maximum frequency of 2.5 GHz can be achieved by this architecture, but at a very significant cost in power consumption. In this case the power consumption increased to ∼12.2× and ∼9.2× compared to 1 GHz DM^2 for the 32- and 64-bit designs, respectively.

The purpose of the second experiment was to compare the DM^2 adder to the RCA at 1 GHz. Unfortunately, the ripple carry adder failed to meet this design goal with the 40 nm technology (due to STD cell library sizing factors). The maximum achievable frequencies were 370 and 195 MHz for the 32- and 64-bit designs, respectively.

Nevertheless, to show that DM^2 is more power efficient than the ripple carry adder, it was optimally designed to meet the 370 MHz and 195 MHz frequencies (32 and 64 bits). In this case the DM^2 design achieved a power reduction of ∼4× (410/102) and ∼5.8× (435/75), compared to the ripple carry adders. It is worth

Table 6.1 Performance, power, area, and number of cells. Comparison of Brent–Kung and ripple adders to DM^2

#bits		Brent Kung	Brent Kung	DM^2	DM^2	RCA
32-bit adder	Power (μW)	410	102	230	593	2840
	Frequency (GHz)	0.38	0.3728	1	1.09	2.503
	Area (μm)^2	488	290	400	824	824
	#Cells	130			435	435
64-bit adder	Power (μW)	435	75	520	1082	4800
	Frequency (GHz)	0.199	0.194	1	1.06	2.142
	Area (μm^2)	967	350	800	1200	1594
	#Cells	260			926	926

Table 6.2 32- and 64-bit control circuitry's (mode decision) average power consumption and performance

Clock cycle (ps)	Delay time (ps)	Av. power consumption (μW)	Control circuit #bits
32	61	320	1000
64	119	380	1000

noting that in this case, the difference in power improvement between the 32- and the 64-bit designs (4 and 5.8) was not significant since some of the DM^2 gates were already minimum-sized given the relaxed performance specifications.

6.1.4.4 Mode Decision Overhead

As discussed above, the mode decision operates in the ID Stage. In order to fully grasp the system tradeoffs, we extracted both the 32- and the 64-bit control circuitry's (mode decision) average power consumption and performance. These are listed in Table 6.2.

As emerges clearly from the table, the mode decision logic delay was much shorter than the clock periods. As shown in Fig. 6.2, the logic used the output of the register file, which is usually not a critical path, and therefore consumed less than half a clock cycle (1 GHz). The ID stage could thus tolerate the incorporation of the decision logic with no timing problems. The simulation results showed that the average power of the mode decision circuitry was ∼20–25% of the DM^2 adder. Up to now, the mode decision average power has not been taken into account (Table 6.1). Although the mode decision clearly injects power overhead, it was negligible and did not disconfirm the advantages of the DM^2 compared to the other alternatives. Table 6.3 presents the average power dissipation of the DM^2 adder including the mode decision unit. As shown, there was a ∼2× (593/291) and ∼1.7× (1082/639) power reduction compared to the Brent–Kung operating at 1 GHz for the 32- and 64-bit designs, respectively. Compared to RCA operating at its maximum frequency, a ∼2.51× (410/163) and ∼2.24× (435/194) power reduction was achieved for the 32- and 64-bit designs.

Table 6.3 Average power dissipation of the DM^2 adder including the mode decision unit

	Brent Kung	DM^2	DM^2	RCA
32-bit power (μW)	410	163	291	593
32-bit frequency (GHz)	0.38	0.3728	1	1.09
64-bit power (μW)	435	194	639	1082
64-bit frequency (GHz)	0.199	0.194	1	1.06

Table 6.4 Computed vs. measured parameter comparison

#bits		k	$\frac{E^{DM^2}}{E^{DMADD}}$ Eq. (6.5)	$\frac{E^{DM^2}}{E^{DMADD}}$ Eq. (6.6)
32-bit adder	Computed	12.2	0.59	0.59
	Measured	8	0.64	
	Inaccuracy [%]		7.8	7.8
64-bit adder	Computed	9.3	0.69	0.65
	Measured	8	0.73	
	Inaccuracy [%]		5.5	11

6.1.4.5 Design Accuracy Analysis

To grasp the accuracy of the optimal DM^2 design analysis, the calculated energy reduction was compared to the SPICE simulation results for 32-bit and 64-bit adders. For the 32-bit adder the following parameters were measured. Note that these parameters are the delays and energy measurements per bit.

$$T_{pre} = 10^{-10} \text{ [s]}, \qquad E_{dyn}^{DM^2} = 4.76 \cdot 10^{-14} \text{ [J]},$$
$$T_{eval} = 3.13 \cdot 10^{-11} \text{ [s]}, \qquad E_{stat}^{DM^2} = 6.42 \cdot 10^{-15} \text{ [J]}, \qquad (6.9)$$
$$T_{stat} = 4.68 \cdot 10^{-11} \text{ [s]}, \qquad E_{stat}^{DM^2} + E_{leakage}^{DMO} = 1.08 \cdot 10^{-14} \text{[J]}.$$

For the 64-bit adder the following parameters were measured:

$$T_{pre} = 10^{-10} \text{ [s]}, \qquad E_{dyn}^{DM^2} = 4.96 \cdot 10^{-14} \text{ [J]},$$
$$T_{eval} = 1.56 \cdot 10^{-11} \text{ [s]}, \qquad E_{stat}^{DM^2} = 7.24 \cdot 10^{-15} \text{ [J]}, \qquad (6.10)$$
$$T_{stat} = 6.25 \cdot 10^{-11} \text{ [s]}, \qquad E_{stat}^{DM^2} + E_{leakage}^{DMO} = 1.11 \cdot 10^{-14} \text{ [J]}.$$

Table 6.4 shows the measured k derived from Figs. 6.7 and 6.8 for the 32-bit and 64-bit adders, respectively, and the corresponding rounded off k (for practical and obvious reasons). Energy reductions of 36% and 27%, respectively, were achieved. Note that the computed energies are lower bounds since the group size k may be smaller in practice than the computed k due to rounding off.

The table shows that the energies measured in the simulations are close to those computed by Eq. (6.5), with only small inaccuracies of 7.8% and 5.5%, respectively, for the 32-bit and 64-bit adder designs.

6.1.4.6 Reliability

Dynamic voltage and frequency scaling (DVFS) has become a highly popular energy reduction technique. Sensitivity to process variations has also become a major design concern. This explains why it is so important to verify the voltage scalability of the DM^2 design and its sensitivity to process variations. Ideally, we want the minimum energy design point at which the group size k was determined to be invariant to the operation voltage.

Recall that the value of k is set such that the n-bit DML dynamic mode propagation delay is equal to a $(2k - 1)$-bit static mode propagation delay. This means that the delay ratio should be independent of the operation voltage. The following expression shows an approximate delay ratio:

$$
\frac{T_{eval(n)}}{T_{stat(2k-1)}} = \frac{nC_{dyn} \int_0^{V_{DD}/2} \frac{dV}{I_{dyn}}}{(2k-1)C_{stat} \int_0^{V_{DD}/2} \frac{dV}{I_{stat}}}
$$

$$
= \frac{\frac{nC_{dyn}}{\delta_{dyn}} \int_0^{V_{DD}/2} \frac{dV}{f(V_{DD})}}{\frac{(2k-1)C_{stat}}{\delta_{stat}} \int_0^{V_{DD}/2} \frac{dV}{f(V_{DD})}} \approx \frac{nC_{dyn}\delta_{stat}}{(2k-1)C_{stat}\delta_{dyn}}
$$

(6.11)

The constant ratio in Eq. (6.11) stems from the current equation $I_{dyn} = \delta_{dyn} f(V_{DD})$ and $I_{stat} = \delta_{stat} f(V_{DD})$. The factors δ_{dyn} and δ_{stat} are the current driving strength of the respective topologies, which depend solely on the device sizes and process parameters. $f(V_{DD})$ depicts the current dependency on the supply voltage, when the device is operated in one of the possible operation modes, e.g., strong inversion, near-threshold, and subthreshold.

Figure 6.12a illustrates the n-bit dynamic and $(2k - 1)$-bit static delays in a logarithmic scale. The two curves should theoretically coincide. They may in actuality be separated slightly as a result of rounding off (see Fig. 6.7). Figure 6.12b depicts the ratio of these delays and shows that it is almost constant across a wide voltage range.

To examine its sensitivity to process variations, the DM^2 and ordinary CMOS DMADD adders were tested by running 2000 Monte-Carlo simulations for its static and dynamic modes. The results are summarized in Table 6.5 and show only a very small change in the sensitivity of the DM^2 adder compared to the DMADD. This comes as no surprise, since DML was previously shown to be robust.

6.1.4.7 Area Utilization

DM^2 and DMADD adders were designed to compare their areas. Figure 6.9 shows the layout of the DM^2 which was custom-designed. The DMADD was synthesized with a Cadence Encounter RTL Compiler. The DM^2 was 32% smaller than the

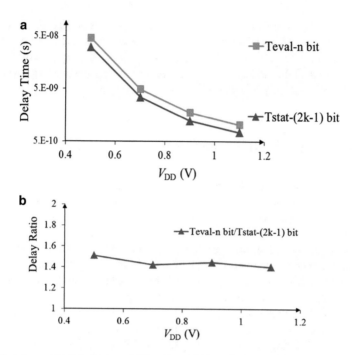

Fig. 6.12 (**a**) $n = 64$-bit dynamic and $(2k - 1) = 16$-bit static delays and (**b**) their ratio

Table 6.5 2000 runs Monte-Carlo delay results

	Dynamic 64-bit	Static 16-bit	DMADD CMOS 64-bit
Delay variance (σ)	131 ps	58 ps	123 ps

DMADD due to the smaller cell sizes of the DML family compared to the CMOS (in DML, either the pull-up or the pull-down transistor network is always the minimum size).

6.1.5 Coarse-Grain Control Conclusions

In this subsection we demonstrated how the dual-modularity of the DML gates, along with architectural and block-level control signals, can be exploited to enable considerable gains in performance, energy, and area. The mode selection was done in coarse-grain for a complete logical block. The test case discussed was a low-energy, high-performance DM^2 adder combining DML logic and dual-mode addition to control the modes. This simplifies the usage of dual-mode addition in a pipelined processor while further reducing the computation energy by 36–27% for 32-bit and 64-bit adders, respectively, compared to the DMADD implementation. The DM^2 adder required 32% less area and its robustness for process variations was clearly proven. The combination of novel circuit topologies and a probability-based computational circuit architecture thus has the potential to achieve considerably higher efficiency than traditional designs. We believe that specifically tailored solutions (like the DM^2 adder) can be employed for multipliers and larger arithmetic circuits that are currently being used in electronic systems.

6.2 Fine-Grain DML Mode Selection Controller

In this subsection the goal is to present a much finer-grain selection of DML modes. That is, we will be looking at mode selection per logic path in a circuit in compliance with system requirements. For this purpose we take a Carry Look-Ahead adder as an example.

Fast data processing abilities are highly influenced by the ALU's implemented processor adder speed. Statistically, addition operations are often the main instruction performed while processing [28, 29]. Many adder architectures have been researched, analyzed, and proposed for speed improvement and power reduction [30]. The Carry Lookahead Adder (CLA) [31, 32] was suggested as an alternative for the speed enhancement of a simple ripple adder [31]. CLA speed is usually determined by the slowest critical carry path delay. In general, the CLA critical path is data-dependent and changes during CLA operation. Current solutions improve the slowest critical path to ensure proper operation in the worst case. Typically, these solutions improve the slowest CLA critical path delay by a sizing optimization of the CMOS gates or implementation with alternative design styles such as PTL or dynamic logic [30, 33]. This speed improvement is however associated with a significant increase in the power dissipation of the adder.

The CLA discussed here was implemented with dual mode logic (DML) utilizing the very low power dissipation of the static mode at moderate performance and higher performance in the dynamic mode, although with increased power dissipation. The CLA utilizes this powerful capability of DML by a dynamic selection of critical paths according to the input vectors. The critical paths (and only those) are operated in the dynamic mode and improve the CLA delay. The remainder of the CLA operates in the DML static mode, thus improving CLA power consumption. A 32-bit DML CLA was designed in a 40 nm low-power TSMC process. Simulation results showed a 45% gain in speed and 70% in power dissipation compared to the CMOS and dynamic CLA, respectively. In addition, the simulations evidence full functionality and robustness to global and local process variations at supply voltages as low as 0.6 V.

Below, we review the characteristics of DML and then present the CLA design approach, including the CLA architecture and principles of operation. We end with a performance analysis of the CLA and summarize the results.

6.2.1 Design Example: Carry Look-Ahead Adder

We start this subsection by underscoring several DML-related design considerations which need addressing when we want to employ logic path-based DML mode selection. Although design with DML methodology is generally very intuitive and has been discussed in detail in the previous chapters, there are a few important points related to the design of the CLA that need to be emphasized:

1. Implementation through NP-like blocks: In general, when cascading dynamic logic gates (such as dynamic-NP or domino gates), the family structure imposes several limitations. The design here was implemented through un-footed DML gates, which were cascaded in an NP-like fashion.
2. If an un-footed DML is connected in a cascade, the inputs to the DML *Type-A* gate must come from the DML *Type-B* gate and vice versa. This rule aims to prevent short circuit currents and a possible fault mechanism. In particular, in complex systems, this specification requires special design solutions.
3. When designing a standalone un-footed DML system dedicated for use in a CMOS-like environment (or between registers), the inputs to the system during the precharge must be kept high or low, depending on whether the system is implemented using *Type-A* or *Type-B*. We solve this issue by implementing the footer solely at the first input-chain gate.

The CLA design enables shorter delays at the cost of higher hardware complexity [29]. In the standard CMOS implementation, the critical path will always be the longest carry route and is determined by the number of bits: $2log_2(N) - 1$. The solution here allows the critical path of the CLA to be selected dynamically and accelerated by operating the DML gates of that path in the dynamic mode. The critical path is identified according to the inputs during operation per clock cycle and set by $2log2(i) - 1$, where $i < N$ is the max index of the generated carry. The longer the critical path, the greater the improvement in delay. To implement this mechanism, decision logic, which identifies the critical path and creates the appropriate clock signals for the DML gates in this path, is required.

6.2.1.1 Decision Logic

The decision whether to accelerate a carry route is based on the input to the CLA. This route should only be dynamically operated in cases where the carry-out is generated. Table 6.6 depicts a simple case where two corresponding bits X_i and Y_i of CLA inputs X and Y are examined. It shows that a simple NOR operation between X_i and Y_i will ensure the dynamic operation of the path in the case of carry-out generation. However, this simple solution will also operate 33% of the paths dynamically when the dynamic operation is not required, resulting in increased power dissipation of the adder. Figure 6.13 illustrates an implementation of a simple single-bit decision circuitry which is responsible for switching between the dynamic

Table 6.6 False–true dynamic activation as a function of the input vectors

X_i	Y_i	C_{in}	C_{next}	Dynamic activation is needed	$NOR(x_i, y_i)$: when $= 0$ the route is dynamically activated	False dynamic activation
0	0	0	0		1	
0	0	1	0		1	
1	0	0	0		0	Yes
1	0	1	1	Yes	0	
0	1	0	0		0	Yes
0	1	1	1	Yes	0	
1	1	0	1	Yes	0	
1	1	1	1	Yes	0	

Fig. 6.13 Single-bit mode-decision. The PTL is composed of low-V_T devices

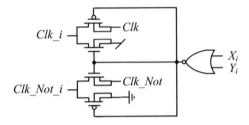

and static modes of operation. This very simple single-bit decision circuitry consists of two low-V_T based transmission gates controlled by a two-input NOR gate.

The decision is made based on X_i and Y_i bits (one bit from each input), and the circuits' outputs are connected to the precharge transistors of the DML logic on the carry route. The CLK_i controls the precharge operation of *Type-A* DML logic, and CLK_Not_i controls the precharge of the *Type-B* logic. In situations where dynamic operation is not required, the outputs of the decision circuitry disable all M_1 transistors in the route, resulting in the static operation of the DML gates. The system is self-controlled and self-switched between static and partial-dynamic operations.

A very simple single-bit decision circuit is illustrated. However, a more precise decision can be implemented using more complex k-bits based decision gates, which will decrease the power dissipation by reducing the number of dynamically operated gates at the expense of area. It can be shown that the optimal solution (via a power–area tradeoff) can be achieved with a 2–4-bit based decision circuitry.

6.2.1.2 CLA Architecture

The architecture of the 32-bit CLA is shown in Fig. 6.14. The core of the CLA is very similar to the well-known highly investigated conventional CLA design composed of two basic building blocks: A and B. The functionality of these blocks is as follows:

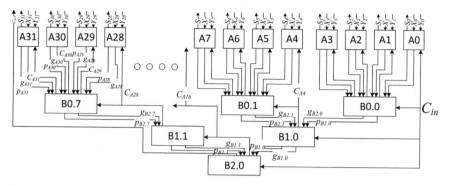

Fig. 6.14 Improved CLA implementation

The B blocks are indexed by m, j, where m represents the hierarchical level of the block ($m = 0, 1, 2$). The inputs and outputs to a B block are presented in Fig. 6.15a. As can be seen in Fig. 6.15b, the X and Y indices are in the form of $i + k_l$ (where l is a subseries, $l = 0, 1, 2, 3$). The i, k indexes depend on the B block hierarchical level, such that they depend on m ($m = 0, 1$ or 2):

$$
\begin{aligned}
m = 0 &\rightarrow i = 0, 4, 8, \ldots, 4n; \, k_l = 1, 2, 3; \\
m = 1 &\rightarrow i = 3, 19, 35, \ldots, (4^2 n + 3); \, k_l = 4, 8, 12; \\
m = 2 &\rightarrow i = 15, 79, \ldots, (4^3 n + 15); \, k_l = 16, 32, 48.
\end{aligned}
\tag{6.12}
$$

Inspection of the $B_{0,j} (m = 0)$ block shown in Fig. 6.15 reveals that besides the standard structure and logic functions which exist in conventional CMOS implementations, there are four extra single-bit decision circuits. Each circuit is responsible for the dynamic or static operation of a specific path. For example, if the X_i, Y_i inputs are such that a carry is needed, the $C_{A(i+1)}$ route is dynamically operated. Note that if inputs X_{i+3}, Y_{i+3}; $Y = 0, 4, 8, \ldots, 28$ are such that a carry is needed, the carry is the output of a higher hierarchical level B block ($B_{m,j}$), and therefore the entire carry route ($C_{A(i+4)}$; $Y = 0, 4, 8, \ldots, 28$) is dynamically operated. This means that the inputs p_{out}, g_{out} (or $pB(2, i \bmod 3)$, $gB(2, i \bmod 3)$) are dynamically operated from the $B_{0,j}$ block (Table 6.7).

It is clear that this recursive simple structure can be expanded to any CLA size. The third level four-bit CLA is a recursive implementation of the previous level which only uses half of its hardware.

It is important to note that gates with a footer should be used at the first level of each B block to allow for efficient precharge. To enable correct operation, static signals must be stable at the system's inputs before evaluation, i.e., they must overlap the precharge periods. This can be done by pipelining, which is not discussed here.

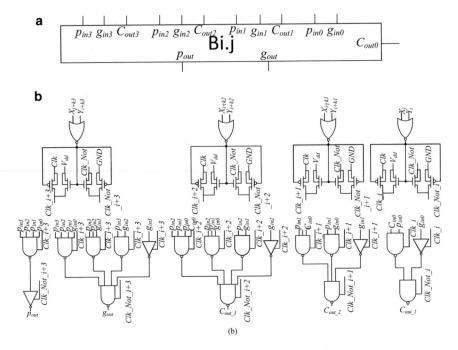

Fig. 6.15 (**a**) Inputs and outputs of a B block and (**b**) the structure of the $B_{0,j}$ block

Table 6.7 A and B block functionalities

Block A	Block B
$s_i = X_i \oplus Y_i \oplus C_{Ai}$	$P_{out} = \prod\limits_{r=0}^{3} p_{in_r}$
$g_{Ai} = X_i Y_i$	$g_{out} = g_{in0} p_{in1} p_{in2} p_{in3} + g_{in1} p_{in2} p_{in3} + g_{in2} p_{in3} + g_{in3}$
$p_{Ai} = X_i + Y_i$	$C_{out1} = g_{in0} + p_{in0} C_{in}; C_{out\_2} = g_{in1} + g_{in0} p_{in1} + C_{in} p_{in0} p_{in1}$
	$C_{out3} = g_{in2} + g_{in1} p_{in2} + g_{in0} p_{in1} p_{in2} + C_{in} p_{in0} p_{in1} p_{in2}$

6.2.2 Fine-Grain Controller Simulation Results

The 32-bit CLA was tested and characterized in a low power 40 nm TSMC process using a SPICE-based Virtuoso simulator. Power supplies between 0.6 and 1 V were applied to examine proper functionality. CLA functionality was examined in the presence of global and local process variations. DML CLA performance, power dissipation, and area were compared to the CMOS and dynamic counterparts. Figure 6.16 depicts an example of the simulation of two routes ($out<3>$ and $out<31>$) and the global clock under standard 1.1 V operation. As shown, the routes are operated dynamically or statically, depending on the input (which are not shown in Fig. 6.16). As can be seen, for example, the route of $out<31>$ is in the dynamic mode at 2.9 µs, whereas the route of $out<3>$ is computed statically.

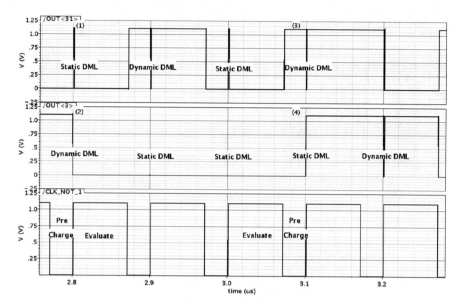

Fig. 6.16 Transient analysis of two routes

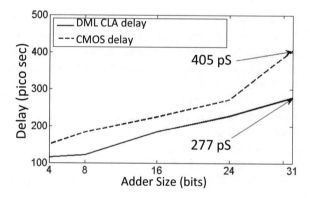

Fig. 6.17 DML CLA delay compared to CMOS CLA

While the DML CLA achieves the same performance as a fully dynamically operated adder (the proposed architecture ensures the dynamic operation of all critical paths), it exhibits improved delay compared to the CMOS CLA. Figure 6.17 compares the delay of the DML CLA to the conventional CMOS implementation. The delay is shown as a function of the adder size. As illustrated, the proposed architecture can achieve a delay improvement of up to 45% for a 32-bit size adder.

Energy dissipation (for a single computation) of the DML CLA architecture versus CMOS and the fully dynamic architecture is shown in Fig. 6.18.

The energy was measured for the case with simple input vectors (a small number of dynamic carry routes) and for the case with complex vectors (eight paths of

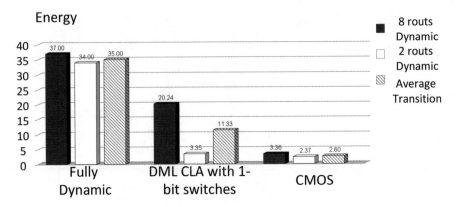

Fig. 6.18 Energy per transition comparison

Table 6.8 Transistor count comparison

Table content: transistor count	CMOS	DML CLA with single-bit switches
32-bit adder	2524	3834
64-bit adder	5180	7882

the design are dynamically operated). As can be seen, the DML CLA architecture achieved significant power reduction compared to the fully dynamic logic CLA. On the other hand, the fully static implementation had lower power dissipation. The power dissipation of the DML CLA increased as the vector became more complex (eight concurrent dynamic paths).

Table 6.8 compares the number of transistors in CMOS and DML of 32- and 64-bit adders. The area overhead decreases for large adders, where the overhead due to additional switches becomes negligible, while the precharge and footer device overhead tends toward a constant value.

Note that a larger number of transistors does not necessarily imply a larger layout or capacitances since about 50% of the transistors in the DML implementation are minimum-sized.

6.2.3   Fine-Grain Control Conclusions

This subsection described how the dual modularity of the DML gates can be utilized along with architectural and local path-level control signals to achieve considerable gains in performance with minimum energy overhead. The mode selection involved a fine-grain per critical path in the design and was very local. This approach yielded high-performance CLA. By building the CLA while using DML logic, we provide a way for the critical path of the CLA to be dynamically chosen and accelerated by operation in the dynamic mode. The architecture of the DML CLA was presented as well as its operating principles.

References

1. W. Kim, M.S. Gupta, G.-Y. Wei, D. Brooks, System level analysis of fast, per-core DVFS using on-chip switching regulators, in *2008 IEEE 14th International Symposium on High Performance Computer Architecture* (IEEE, Piscataway, 2008), pp. 123–134
2. B.R. Zeydel, D. Baran, V.G. Oklobdzija, Energy-efficient design methodologies: high-performance VLSI adders. IEEE J. Solid State Circuits 45(6), 1220–1233 (2010)
3. H.Q. Dao, B.R. Zeydel, V.G. Oklobdzija, Energy optimization of pipelined digital systems using circuit sizing and supply scaling. IEEE Trans. Very Large Scale Integr. Syst. 14(2), 122 (2006)
4. W. Shen, Y. Cai, X. Hong, J. Hu, An effective gated clock tree design based on activity and register aware placement.IEEE Trans. Very Large Scale Integr. Syst. 18(12), 1639–1648 (2010)
5. J. Shinde, S. Salankar, Clock gating—a power optimizing technique for VLSI circuits, in *2011 Annual IEEE India Conference* (IEEE, Piscataway, 2011), pp. 1–4
6. K. Roy, S.C. Prasad, *Low-Power CMOS VLSI Circuit Design* (Wiley, London, 2009)
7. M. Alioto, Ultra-low power VLSI circuit design demystified and explained: a tutorial. IEEE Trans. Circuits Syst. I: Reg. Papers 59(1), 3–29 (2012)
8. D. Bol et al., Robust and energy-efficient ultra-low-voltage circuit design under timing constraints in 65/45 nm CMOS. J. Low Power Electron. Appl. 1(1), 1–19 (2011)
9. H. Zhang, J. Rabaey, Low-swing interconnect interface circuits, in *Proceedings of the 1998 International Symposium on Low power Electronics and Design* (ACM, New York, 1998), pp. 161–166
10. J.-S. Seo, H. Kaul, R. Krishnamurthy, D. Sylvester, D. Blaauw, A robust edge encoding technique for energy-efficient multi-cycle interconnect. IEEE Trans. Very Large Scale Integr. Syst. 19(2), 264–273 (2011)
11. S.J. Wilton, S.-S. Ang, W. Luk, The impact of pipelining on energy per operation in field-programmable gate arrays, in *International Conference on Field Programmable Logic and Applications* (Springer, Berlin, 2004), pp. 719–728
12. S. Kiamehr, M. Ebrahimi, M.S. Golanbari, M.B. Tahoori, Temperature-aware dynamic voltage scaling to improve energy efficiency of near-threshold computing. IEEE Trans. Very Large Scale Integr. Syst. 25(7), 2017–2026 (2017)
13. S. Höppner, Y. Yan, B. Vogginger, A. Dixius, J. Partzsch, F. Neumärker, S. Hartmann, S. Schiefer, S. Scholze, G. Ellguth et al., Dynamic voltage and frequency scaling for neuromorphic many-core systems, in *2017 IEEE International Symposium on Circuits and Systems (ISCAS)* (IEEE, Piscataway, 2017), pp. 1–4
14. F. ur Rahman, V. Sathe, Quasi-resonant clocking: continuous voltage-frequency scalable resonant clocking system for dynamic voltage-frequency scaling systems. IEEE J. Solid State Circuits 53(3), 924–935 (2018)
15. S.B. Nasir, S. Sen, A. Raychowdhury, Switched-mode-control based hybrid LDO for fine-grain power management of digital load circuits. IEEE J. Solid State Circuits 53(2), 569–581 (2017)
16. S. Bang, W. Lim, C. Augustine, A. Malavasi, M. Khellah, J. Tschanz, V. De, 25.1 a fully synthesizable distributed and scalable all-digital LDO in 10 nm CMOS, in *2020 IEEE International Solid-State Circuits Conference-(ISSCC)* (IEEE, Piscataway, 2020), pp. 380–382
17. F. Atallah, K. Bowman, H. Nguyen, J. Jeong, D. Yingling, Y. Sun, B. Appel, A. Polomik, M. Harinath, J. Morelli et al., 19.3 a 7 nm all-digital unified voltage and frequency regulator based on a high-bandwidth 2-phase buck converter with package inductors, in *2019 IEEE International Solid-State Circuits Conference-(ISSCC)* (IEEE, Piscataway, 2019), pp. 316–318
18. S. Wimer, A. Albeck, I. Koren, A low energy dual-mode adder. Comput. Electr. Eng. 40(5), 1524–1537 (2014)
19. A. Kaizerman, S. Fisher, A. Fish, Subthreshold dual mode logic. IEEE Trans. Very Large Scale Integr. Syst. 21(5), 979–983 (2013)
20. I. Levi, A. Belenky, A. Fish, Logical effort for CMOS-based dual mode logic gates. IEEE Trans. Very Large Scale Integr. Syst. 22(5), 1042–1053 (2014)

21. I. Levi, A. Fish, Dual mode logic—design for energy efficiency and high performance. IEEE Access **1**, 258–265 (2013)
22. I. Levi, O. Bass, A. Kaizerman, A. Belenky, A. Fish, High speed dual mode logic carry look ahead adder, in *2012 IEEE International Symposium on Circuits and Systems* (IEEE, Piscataway, 2012), pp. 3037–3040
23. I. Levi, A. Kaizerman, A. Fish, Low voltage dual mode logic: model analysis and parameter extraction. Microelectron. J. **44**(6), 553–560 (2013)
24. P. Behrooz, *Computer Arithmetic: Algorithms and Hardware Designs*, vol. 19 (Oxford University Press, 2000), pp. 512583–512585
25. K.C. Yeager, The Mips R10000 superscalar microprocessor. IEEE Micro **16**(2), 28–41 (1996)
26. N.H.E. Weste, D.M. Harris, *CMOS VLSI Design: A Circuit and System Perspective*, 4th edn. (Pearson Education India, 2015)
27. R.P. Brent, H.-T. Kung, A regular layout for parallel adders. IEEE Trans. Comput. **3**, 260–264 (1982)
28. J.L. Hennessy, D.A. Patterson, *Computer Architecture: A Quantitative Approach* (Elsevier, Amsterdam, 2011)
29. M.A. Franklin, T. Pan, Performance comparison of asynchronous adders, in *Proceedings of 1994 IEEE Symposium on Advanced Research in Asynchronous Circuits and Systems* (IEEE, 1994)
30. F.-C. Cheng, S.H. Unger, M. Theobald, Self-timed carry-lookahead adders. IEEE Trans. Comput. **49**(7), 659–672 (2000)
31. C.R. Kime, M. Morris Mano, *Logic and computer design fundamentals* (Prentice Hall, 2003)
32. I. Flores, *The logic of computer arithmetic* (1963)
33. A. De Gloria, M. Olivieri, Statistical carry lookahead adders. IEEE Trans. Comput. **45**(3), 340–347 (1996)

Chapter 7
Towards a DML Library Characterization and Design with Standard Flow

After discussing the DML foundations and presenting several conceptual use cases of DML, we now introduce the reader to ways to scale up the utilization space of DML. Specifically, this chapter presents an approach to a DML cell-library characterization and describes the methodology that paves the way for the design of DML circuits using standard tools. We detail the specific library characterization process and show how to design DML systems using standard Electronic Design Automation (EDA) tools. Finally, we synthesize the results from a large number of benchmarks and compare them to standard CMOS flow. While the results indicate that DML design flow can enable exploitation of DML advantages, they also show that standard flow still cannot really exploit the inherent advantages of DML and still does not provide a solid solution for DML-based designs. In the next chapter we discuss a different alternative to utilizing standard design tools which shows that there are many other ways to better harness DML's unique benefits.

7.1 Introduction

In the previous chapters we looked at several custom designs of DML circuits that have been shown to be very efficient. The goal of this chapter is to demonstrate that the DML logic family can be compatible with the standard design flow and be optimized by various tools such as synthesis and physical design in the future. Nevertheless, implementing DML circuits using the standard design flow and EDA tools is highly challenging, since DML gates operate in two different modes, each with its own characteristics and operating mechanisms. We take the example of one of the possible ways to integrate DML with existing and conventional design tools.

Standard static-logic compatible EDA flows have received empirical attention for almost 50 years. Multiple algorithms and heuristics have been put forward and tested for each step and tool of the design flow, and numerous abstractions have been added to reduce complexity [1–3]. These EDA tools have reached a high level of

© Springer Nature Switzerland AG 2021
I. Levi, A. Fish, *Dual Mode Logic*, https://doi.org/10.1007/978-3-030-40786-5_7

integrity and have yielded cross-verified quality results with a short time-to-market. However, fundamental design challenges still arise when the logic underpinning the automation tools is dynamic-based [4–8].

In this chapter we introduce a design flow for DML that overcomes a few of the best-known dynamic design hurdles. This automation method is implemented on a dedicated DML standard cell library that was constructed and fully characterized for this book. The DML flow offers fully automation-compatible steps, innovative approaches, and the use of standard tools, which results in a flexible design in terms of energy–delay tradeoffs that capitalize on the inherent capability of the DML to switch between different operational modes. The main points discussed as follows:

- The formulation of a DML standard cell-library characterization methodology and the ability to utilize it within a design flow.
- The adaptation of the standard design flow (STDF) to DML when utilizing popular commercial EDA tools to address the unique needs of DML.

The structure of this chapter is as follows. We start by presenting the challenges involved in DML adaptation to the automated application-specific integrated circuit (ASIC) design flow and then shift to the digital design flow for DML and the DML library characterization methodology. We close this chapter by summarizing the outcomes of this DML standard library characterization and EDA flow with practical results.

7.2   Characterization and Standard Design Flow Challenges

This subsection presents some of the challenges associated with a DML cell-library characterization and digital flow integration.

7.2.1   Standard Design Flow: Overview and DML Integration Challenges

For a number of reasons, automated ASIC design flows have been governed by the need to simplify and abstract the underlying models of cells and primitives. Satisfactory reliable abstraction relies on the characteristics of the CMOS digital logic family and in particular its superior robustness and large noise margins, rail-to-rail logic levels, unidirectional information flow, and low leakage. Nowadays, the key virtue of static CMOS is its design compatibility with EDA tools, which continue to evolve considerably, since changes in the logic family can lead to unexpected and deteriorated results from EDA tools. When using CMOS, a relatively small set of design rules must be applied for successful EDA integration. This

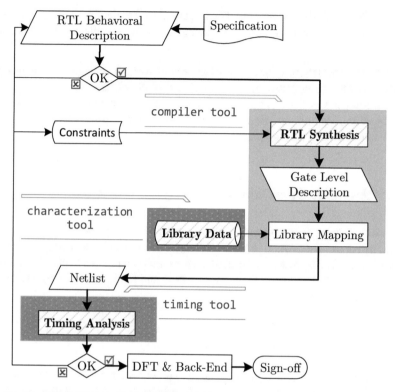

Fig. 7.1 Simplified ASIC standard design flow

clearly reduces the automation tool complexity. Needless to say, most contemporary standard EDA tools are also completely oriented toward static CMOS designs. The ASIC industry adheres to a well-defined and tested STDF throughout the entire design cycle, starting with the product specification definition up to submission of the production files to the foundry. A typical flow fragment of the design phase (verification excluded) is shown in Fig. 7.1 [3]:

The key STDF stages are:

1. RTL-synthesis: conversion of RTL to generic gates and registers while optimizing the logic efficiency and then mapping it to a real library of characterized cells by revamping it for best design metrics.
2. Standard cell library: a library of real, laid-out, and characterized logic gates for the synthesizer mapping process. These include the geometric, timing, and power metric data.
3. Static Tinimg Analysis (STA): Evaluation of all timing paths within the logic networks of the design and monitoring for any timing constraint violations.

For simplicity, a DML gate can be depicted as a static CMOS gate that can operate dynamically when an active clock is applied to it. By contrast, an inactive clock signal will degenerate the DML gate into its CMOS counterpart. This abstraction cannot be exploited as a starting point for a static STDF adaptation. However:

1. DML operates statically or dynamically, and as such, the design flow needs to be able to differentiate between these two substantially different cases.
2. As introduced in Sect. 7.2.2.1, the composition of a dynamic logic network is subject to a bipartite criterion which implies that not all strictly valid designs are feasible from a dynamic operation perspective.

7.2.2 Dynamic Operation Mode Design Challenges

This subsection highlights the main challenges in the design of the DML dynamic mode (where standard dynamic logic challenges are, generally speaking, a subset of these obstacles). We present these challenges as a foundation for Sects. 7.3 and 7.4, where the proposed approach to characterization and Design Flow (DF) integration tackles them.

7.2.2.1 Non-unate Boolean Functions

As defined in [1], a function is unate in all its variables if and only if it is either monotonically increasing or monotonically decreasing for all of its variables. Monotonicity in a variable x_1 described as:

$$
\begin{aligned}
f_{inc}(0, x_2 \ldots, x_n) &\leq f_{inc}(1, x_2 \ldots, x_n) \qquad \forall (x_2 \ldots, x_n) \\
f_{dec}(0, x_2 \ldots, x_n) &\geq f_{inc}(1, x_2 \ldots, x_n) \qquad \forall (x_2 \ldots, x_n)
\end{aligned}
\tag{7.1}
$$

Hence, the unateness and monotonicity terms are interchangeable throughout this chapter. The non-unate functions are crucial to dynamic logic-based designs because implementing them can lead to an area increase and the addition of logically redundant gates. In what follows, we clarify this point; specifically, dynamic logic takes advantage of absent or degraded complementary evaluation networks [9–11] and thus propagates faster than CMOS logic. As explained in Chapter 2, all dynamic logic styles must apply a proper cascading policy of evaluation networks to ensure correct data propagation. Figure 7.2a provides an example of improper cascading of dynamic gates and the subsequent corruption of the propagating data. This example makes it clear that right after the precharge (start of evaluation) of serially connected nMOS evaluation network-based dynamic gates, the second gate will start discharging regardless of the actual output to be evaluated at the output of the first gate.

This concept can be generalized even further. If a logic gate with a preferred nMOS evaluation network is seen as a vertically striped vertex (*Type-A* in DML

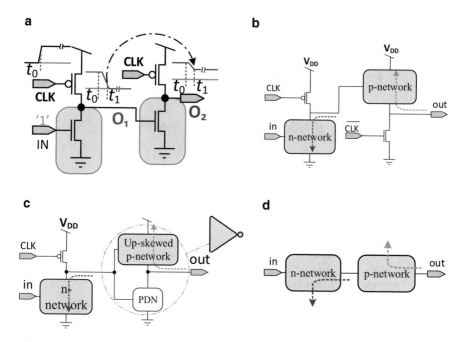

Fig. 7.2 Color classification of dynamic logic cells: (**a**) improper dynamic cascading link, (**b**) np-CMOS (NORA) cascading link, (**c**) n-domino cascading link, and (**d**) generic dynamic cascading link

terminology), and a gate with a pMOS evaluation network is seen as a horizontally striped vertex (*Type-B* in DML terminology), a correct dynamic logic cascading exists if every vertex has complementary predecessor and successor vertices (horizontal and vertical stripes). This concept has its analogy in graph theory and is known as a two-colored graph or bipartite graph [12]. Figure 7.2 depicts the color classification of the (b) np-CMOS, (c) n-domino logic families, and (d) the generic dynamic style.

The output logic level of a non-unate Boolean function $f_{nu}(\bar{x}) = f_{nu}(x_1, x_2 \ldots, x_n)$ will evaluate to logical "0" or "1" depending on the change of x_i and the status of the other inputs $x_{i \neq j}$ [13]. This implies a reconvergence of paths with unbalanced odd and even numbers of logic stages prior to the non-unate stage (gate of reconvergence), G_f. $G_0, G_1 \ldots, G_n$, shown in Fig. 7.3, represent logic levels from the primary inputs to the output $f_{nu}(\bar{x})$. A non-unate function can always be represented as in Fig. 7.3, where the difference in logic levels between two reconvergent branches must be odd.

As depicted in Fig. 7.3, a logic network of non-unate Boolean functions must have at least two reconvergent paths with an odd difference of logic depths. This conflicts with the cascading policy described above, such that none of the CMOS-based dynamic logic styles is able to implement these functions as is. Unfortunately,

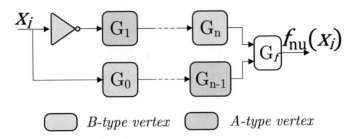

Fig. 7.3 Typical network fraction of a non-unate Boolean function

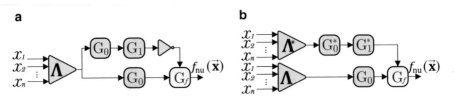

Fig. 7.4 Logic cone duplication and bubble pushing: (a) before and (b) after

these directly inapplicable sets of non-unate Boolean functions include a few vital functions such as XOR or MUX. The only conventional generic method to deal with this restriction, unless a clock signal is also used to control the data flow, is the duplication of the logic cones prior to the non-unate function node, which transforms its logic cone into a monotonic network [1, 14].

Logic duplication is also closely related to the "trapped inverter" problem in domino logic [8], where both polarities of the input signals are required simultaneously but standalone inverters are not available. This method is rescued in part by recursive unate transformations (bubble pushing) before the duplication that push the trapped inversion stages down the logic path to the primary inputs (PI), thus preserving the unateness. The primary inputs of a given dynamic logic domain are defined as its input ports adjacent to other logic domains. Figure 7.4a depicts the general logic structure of a non-unate function, and Fig. 7.4b shows the same function after duplication and bubble pushing, where the $*$ symbol denotes the unate transformed logic and Λ represents a logical cone.

7.2.2.2 Dynamic Operation Characterization Challenges

In this subsection, we briefly present the main characterization challenges facing a DML standard cell library (*.lib*).

1. Characterization of asymmetric behavior: As stated above, DML is capable of operating the gates in the dynamic mode, which runs with synchronization to the clock signal. Hence, some adjustments to the standard static characterization need to be made. In contrast to static CMOS logic, the evaluation of data

throughout a dynamic logic network is asymmetric (i.e., it only performs one *high-to-low* or *low-to-high* transition); thus the assessment of the propagation delay, input capacitance, and dynamic power must be tailored to differentiate and only capture the specific transitions of logic cells alone (typically the evaluation path under each input transition and gate topology).

2. Inter-data timing relations: Typically, the timing closure provided by the static synthesis tools for combinational cells does not incorporate information on ways to handle a synchronizing input (clock signal) to the gates. The dynamic-mode DML characterization requires the definition of a new inter-data timing relationship to address this issue and is presented in later sections.

7.3 A Step Forward with DML Standard Design Flow

This section paints a step-by-step picture of the proposed dynamic DML flow and then details and explains the rationale for each step according to the design flowchart in Fig. 7.5. To aid in visualization, the flowchart in Fig. 7.5 summarizes the entire workflow, including partitioning into characterization and the design flow. Section 7.3.1 charts the rationale behind the use of a dummy pseudo-static library during synthesis and mapping. Section 7.3.2 details the synthesis and mapping of the DF, including its associated complexities, and Sects. 7.3.2.1 and 7.3.2.2 conclude the DML DF with a few key netlist adaptations and STA.

7.3.1 Pseudo-static Library and Multi-library Representation

As discussed above, both of the DML's functional modes must be verified separately. Thus, each mode requires its own library (with the same gates) describing the characteristics induced by the operating mechanism of the mode. A static library is almost identical to a standard CMOS library, whereas a dynamic library is more complex, because it also contains the timing relationships involved in synchronizing the clock signal *vs.* the data pins.

As noted, the standard synthesis tools are static, but the dynamic mode of DML involves clock synchronization. The assertion of a clock signal, which is used to bring the dynamic logic to a predefined initial condition, is called a precharge (or predischarge) value. This preset phase is followed by a logic evaluation phase during which the clock is inactive and the logic state is determined solely by data pins. Recall, that an inactive clock degenerates the DML into its static form; thus, a dynamic DML gate acts identically to a static gate in terms of data propagation during the evaluation phase. The data begin to propagate from a predefined state and are recurrently pulled up and down by the evaluation networks of the gates throughout the logic path (see Fig. 7.6).

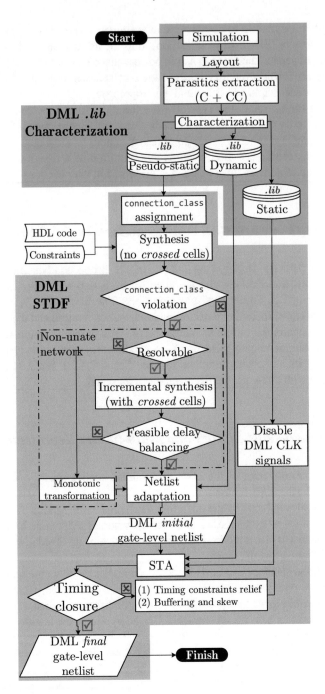

Fig. 7.5 Proposed DML standard design flowchart

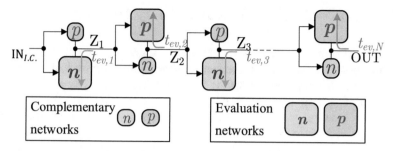

Fig. 7.6 DML data propagation during the dynamic evaluation phase

The arrival evaluation delay at the destination logic OUT node is bounded by the summation value of all transitions on the path (see the left-hand term of the next equation):

$$t_{ev,OUT} = \sum_{i}^{N_n} t_{ev,i}^{(n)} + \sum_{j}^{N_p} t_{ev,j}^{(p)} \Bigg|_{\left\{ \begin{array}{c} t_{ev}^{(n)} = t_{ev}^{(p)} = t_{ev} \\ N_n + N_p = N \end{array} \right\}} = \sum_{i}^{N} t_{ev,i}, \qquad (7.2)$$

where t_{ev} denotes the evaluation phase delay, N is the logic depth of the data propagation path, and the n, p indices denote the PDN and PUN, respectively.

Standard synthesis and timing tools are incapable of computing this evaluation delay. An intuitive way we suggest to manipulate the automatic tool is to duplicate the evaluation network delay (pull-up or pull-down transition) values to the complementary transition of the same gate, which will make the tool take on the correct value regardless of transition direction (as though the gate was static), thus simplifying the timing analysis. For example, a *Type-A* gate always evaluates from *high-to-low*; hence, the characterized (simulated) *high-to-low* delay will be copied to the *low-to-high* delay tables despite the fact that this transition is not possible in dynamic operation. We term this type of library characterization "pseudo-static." This form of characterization can be treated as an enforced symmetric ($t_{ev}^{(n)} = t_{ev}^{(p)}$) dummy pseudo-static .*lib* library of DML cells (as expressed in the previous equation). This library is used for dynamic synthesis when utilizing a static tool. This yields an initial candidate design which is post-processed later on by two additional libraries for complete timing checks of the static and dynamic modes as described in the next subsection.

7.3.2   Pseudo-static Synthesis and Library Mapping

As presented above, dynamic logic networks must enforce a set of rules to ensure their functionality. This set of connection constraints is applied by reading the pseudo-static *.lib* content and setting the `connection_class` attribute [15] of Synopsys® LIBERTY on each of the library cell's data pins. The attribute is considered a design rule and should be followed by the synthesis tool during mapping, since otherwise `connection_class` violations are reported. The mapping process can be enforced to apply these connections. The connection rules of ordinary DML cells defined as in Fig. 7.7 ensure correct cascading.

Note that the LIBERTY `connection_class` attribute support is neglected by the Synopsys® Design Compiler, which was initially utilized for voltage island design. However, it is part of the LIBERTY gate-level library modeling industry standards, which is compatible with all EDA mapping tools if they choose to use it. The `connection_class` is one way to adhere to the connection restrictions. However, it is obvious that designers can implement it differently. Each pin of each library cell is explicitly attributed to a single valid `connection_class` before the mapping, since it is hard for the tool to abide by all the constraints simultaneously if they have been loosely specified. For this purpose, several dummy cells are defined to separate different valid connection scenarios, whereas all the connection classes are defined distinctly at the cell level.

All primary inputs should be imposed with footed cells; otherwise functionality might be lost. This differentiation of PIs is done by assigning a dedicated PI `connection_class`. All the input pins of cells that are authorized to link to these primary inputs are assigned to the PI class. Cells that are valid to link to primary inputs are all footed or semi-footed. Semi-footed refers to a cell that has a serial stack of transistors in the evaluation path; if so, one of its inputs could also interact directly with the primary inputs (as long as the other input can cut off the evaluation network at the beginning of evaluation phase).

As noted above, not all Boolean functions can be realized with a bipartite network; thus the workflow should also be capable of generating correct designs for (a) monotonic and (b) non-unate functions.

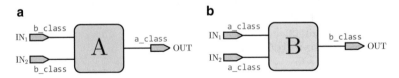

Fig. 7.7 DML regular footless cell connection classes: (**a**) *Type-A* cell and (**b**) *Type-B* cell

7.3.2.1   Monotonic (Unate) Network Mapping

In the case of a monotonic logic network, the mapping process is trivial since no color conflicts are observed. It results in a clean report of `connection_class` violations and valid logic network structures.

7.3.2.2   Non-unate Network Mapping

Not all logic networks can be classified as monotonic, which means that color conflicts are inevitable in the case of a non-unate Boolean function realization. There are two strategies to cope with non-monotonic networks: either making them monotonic [8] or a multi-phased clocking scheme [8, 16]. The latter method is not discussed in this book, since it complicates the design and its clock trees considerably. The monotonic transformation of the logic network is only considered as an alternative solution if the following attempts fail to construct a feasible, valid design.

The report of `connection_class` violations is used to locate the color conflicting nodes of the non-unate logic network. This includes violating the `pin_name` of the conflicting gates and their `cell_name`, which is parsed and is used for the automatic resolution by replacement procedure, where possible.

Before illustrating the removal procedure of `connection_class` violations, we present the cell-naming nomenclature. The output class of the cell is denoted by a capital A or B letter prefix. The class of each IN_i input of the cell is denoted by lowercase a (amber) or b (blue) letters reflected by appending the `cell_name` with suffix letters in their order of appearance. For example, a *Type-A* NAND$_2$ cell is abbreviated A_ND2_bb, and a *Type-A* crossed NAND$_2$ cell written A_ND2_ab describes IN_1 that should be driven by a *Type-A* cell, whereas IN_2 should be driven by a *Type-B* cell. The input connectivity suffix of ordinary DML cells can be omitted, because all the inputs have the opposite connection class to the cell's output. By contrast, crossed cells have a combination of input classes, so more information is required.

Figure 7.8 provides an example of a `connection_class` violation removal procedure, where the conflicting NAND$_2$ cell is replaced by its crossed dummy version. Resolvable conflicting sites are repaired by replacing the conflicting cells with their dummy crossed cell clones. On the other hand, unresolvable color conflicts require a monotonic transformation. Resolvable color conflicts are handled by reinitiating an additional incremental synthesis run, whereas the entire logic network other than `connection_class` violators is frozen for changes (the `dont_touch` attribute is set). In this way, the logic structure remains intact except for the insertion of crossed dummy cells and delay balancing buffers to meet data-to-data timing constraints. The operation of the crossed cells is described in the next paragraph.

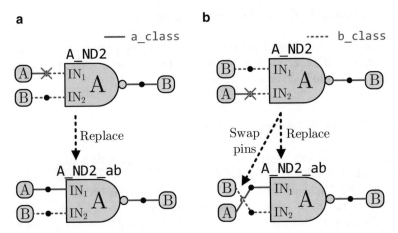

Fig. 7.8 Resolvable color conflicts by crossed cells: (**a**) resolvable color conflict at IN_1 and (**b**) resolvable color conflict at IN_2

Crossed (Stacked) Gates

Some logic gates can absorb two conflicting colors if a specific input condition is met. The basic concept behind these *crossed* cells takes advantage of a serial transistor stack inside the evaluation path. This serial stack allows for a mixture of two signals with opposite precharged states, because one of them cuts off the evaluation path and prevents the gate's false evaluation. Unfortunately, this solution depends on the *data-to-data* timing interdependence of the gate's inputs. Figure 7.9a illustrates this for the case of a *Type-A* (horizontal stripes) NAND$_2$ gate.

Figure 7.9b and c waveforms depict two cases of a cell's correct and false response, respectively. The correct response does not include a transient glitch of output node Z unlike in the failed response, since the input signals do not have a "high"-level overlap. A failed response of the Z output node takes place when both the A and B signals are "high" and trigger an erroneous evaluation. This input data timing criterion imposed by the mapping tool with respect to the *data-to-data* timing constraints is defined in the pseudo-static *.lib*. In the case of a *Type-A* crossed cell, the earliest rise $min(t_{rise}(B))$ of the *Type-B* input should occur later than the latest fall $max(t_{fall}(A))$ of the *Type-A* input. If this condition is met, the *crossed* cell behaves as expected.

Note that a failed response is transient and necessarily converges to its correct value after the arrival of both inputs. However, this incurs a false pull-down and then a pull-up evaluation, which propagates through the rest of the logic nodes and forces them to evaluate the data through their degraded complementary networks, which results in a much slower response. This behavior is unacceptable because it deviates from the modeled timing frame of the entire design and also consumes unnecessary power at the same time.

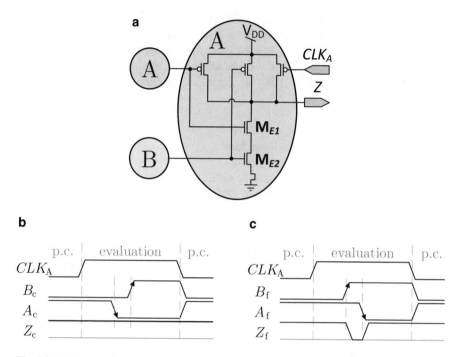

Fig. 7.9 NAND$_2$ crossed cell example: (**a**) crossed *Type-A* NAND$_2$ cell and (**b**) correct evaluation response (**c**) failed evaluation response

Delay Balancing

The delay balancing step assesses the feasibility of meeting the *crossed* cell input data-arrival timing constraint at a reasonable cost (see end of this paragraph). For instance, the automatic insertion of a number of buffers to meet crossed cell inter-data timing might be far more efficient than the duplication of the entire preceding logic cone. If this solution is classified as infeasible, the netlist is recovered in its initial state (no crossed cells or delay balancing). In this book, the criterion of a maximum 10% gate count increase was used. Note that this arbitrary threshold parameter is a choice for the designer to make, based on the design specifications.

Monotonic Transformation

Impractical crossed cell conflict removal leads to the somewhat costly solution of a monotonic transformation of part of the logic network. Unlike the widespread dynamic domino logic design style [9], there is no need for recursive bubble pushing caused by trapped inversion stages [8], since the proposed DML flow is based on discrete inverting logic stages. Thus, the monotonic transformation consists solely of preceding logic cone duplication at conflicting nodes. Logic duplication is carried out with a `monotonic_transform` script that parses the gate-level netlist and replicates the subnetworks prior to the splitting points of the `connection_class` violated cells and re-maps (swap of types) them complementarily (see Fig. 7.4b). The duplication script minimizes logic redundancy by constantly updating and re-using already cloned subnetworks, hence saving on energy and area.

7.3.2.3   Post-Synthesis Netlist Adaptation

At this point, the pseudo-static gate-level netlist is clear of connection rule violations and is ready to be associated with real DML libraries. This is easy to achieve by implementing a few procedural steps:

1. CLK_A and CLK_B signals are added to the global module port list.
2. CLK_A and CLK_B signals are added to the port list of each DML gate locally, and globally propagated to the higher hierarchy via the port list of the module.

   In this way, the full structural DML netlist is ready to be analyzed for timing with a standard STA tool.

7.3.2.4   Static Timing Analysis (STA)

As explained above, the dynamic *.lib* contains various clock-related timing constraints, which need to be verified for proper dynamic functionality. The clock affinity is not included in standard STA tools; hence, the clock port of each DML cell is defined as a dummy data port. These inter-data timing relationships are reminiscent of the conventional setup and entail constraints during the characterization of sequential elements. These timing constraints are satisfied just like standard timing constraints by relaxing the operation frequency and delay corrections which can be remedied with proactive buffering or place and route tools. The STA of a static library has no timing issues by default as long as the operating frequency is met.

7.4   Characterization Process

This section deals with the construction of the multiple standard cell DML libraries required for the proposed DF, as summarized in Fig. 7.5.

The cascading topology of the DML network was chosen to be similar to np-CMOS [11] (see Fig. 7.2b), which utilizes both types (*Type-A* and *Type-B*) of gates and thus gives more optimization space for the synthesis tool. For simplicity and proof of concept, only a small but universal set of logic gates was constructed. This set was composed of $NAND_2$, NOR_2 gates and inverters of both the *A* and *B* types;

Table 7.1 Implemented DML cells with connection classes

Cell name	IN_1 class	IN_2 class	OUT class	Notes
A_INV	Type-B		Type-A	
A_INV_f	PI (Primary input)		Type-A	Footed
A_ND2	Type-B	Type-B	Type-A	
A_ND2_ab	Type-A	Type-B	Type-A	Crossed
A_ND2_pb	PI	Type-B	Type-A	Stacked
A_ND2_f	PI	PI	Type-A	Footed
A_NR2	Type-B	Type-B	Type-A	
A_NR2_sf	PI	Type-B	Type-A	Semi-footed
A_NR2_f	PI	PI	Type-A	Footed
B_INV	Type-A		Type-B	
B_INV_f	PI		Type-B	Footed
B_ND2	Type-A	Type-A	Type-B	
B_ND2_sf	PI	Type-A	Type-B	Semi-footed
B_ND2_f	PI	PI	Type-B	Footed
B_NR2	Type-A	Type-A	Type-B	
B_NR2_ab	Type-A	Type-B	Type-B	Crossed
B_NR2_pa	PI	Type-A	Type-B	Stacked
B_NR2_f	PI	PI	Type-B	Footed

each had several flavors as explained below. Beyond type coding, the DML cells were divided into three additional subcategories:

1. Footless cells: ordinary logic cells that are extensively applied unless a specific condition is encountered.
2. Footed or semi-footed cells: logic cells that have a clock controllable evaluation path. These cells are required for interfacing with other non-DML logic domains. Semi-footed cells have one controllable evaluation path and another ordinary footless evaluation branch. This structure provides a further degree of delay optimization during the mapping process.
3. Dummy cells: auxiliary cells that are logically identical to other footless cells and are used to force the mapping tool to abide by the inter-cell connection rules (see *Type-A* NAND$_2$ example in Fig. 7.8).

Table 7.1 presents the entire set of implemented cells characterized throughout the flow.

Cell Naming Conventions and Types of .libs

`<type>_<cellname>_<connectivity_suffix>`
`<type>` – notes the type of the cell, which implies the precharge value.
`<cell_name>` – abbreviation of logic function and the corresponding fan-in.
`<connectivity_suffix>` – outlines the input connectivity notations:

- f—footed cell.
- sf—semi-footed cell.
- ab—IN_1 has *Type-A* source, while IN_2 has a *Type-B* source.
- pa, pb—IN_1 has PI source, while IN_2 has a *Type-A* or *Type-B* source.

As introduced in Sect. 7.3.1, the DML flow requires three different libraries of characterized cells:

1. Pseudo-static *.lib*: the auxiliary library used for construction of valid logic networks.
2. Dynamic and static *.lib*: real libraries that characterize the design metrics of DML cells in their different operational modes.

The entire set of DML gates was simulated, laid out, and extracted for parasitic elements and represented in the form of a SPICE netlist for further characterization. The following subsections go into the specifics of the characterization process and relate its content to the proposed design automation flow. Note that the characterization of only one type (A) of DML gates is described, since the other type is characterized along the same guidelines as the general procedure.

7.4.1 Pseudo-static Library

The rationale for the pseudo-static library of standard cells is to imitate the dynamic behavior of DML cells without clock synchronization. In addition, it is also subject to the enforcement of connection rules to generate dynamically compatible logic networks.

7.4.1.1 Design Metric Characteristics

To resemble dynamic behavior, cells within the library were characterized solely for relevant evaluation transitions, i.e., input capacitance, intrinsic propagation delay, and power consumption tested during *high-to-low* (*Type-A*) or *low-to-high* (*Type-B*) evaluation transitions. Complementary dummy timing arcs artificially duplicated these characterized metrics:

1. Intrinsic *high-to-low* evaluation delay, measured relative to the data IN port (see Fig. 7.10).
$$i_{ds,n} = C_{Load}\frac{dv}{dt} \Rightarrow t_{ev}^{(n)} = \int_0^{t_{pHL}} dt = C_{Load}\int_{V_{DD}}^{\frac{V_{DD}}{2}} \frac{dv_{out}}{i_{ds,n}(v_{out})}.$$
2. Data input capacitance: $C_{IN} = \frac{\int_0^T i_{IN}(t)dt}{V_{DD}}$.
3. Power consumption: $P_{total} = P_{leak} + P_{dyn}$, where:

    a. $p_{dyn} = \frac{V_{DD}}{T}\int_0^T i_{V_{DD},dyn}(t)dt$, and
    b. $P_{leak} = V_{DD} \cdot I_{V_{DD},leak}$

Fig. 7.10 Intrinsic *high-to-low* evaluation delay definition

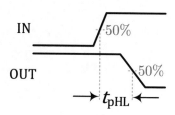

where $I_{V_{DD},leak}$, $i_{V_{DD},dyn}$ are the leakage static current and the dynamic transient current flowing through the cell and P is the stand notation for power.
4. Area footprint was taken from the layout.

Some of these standard cell characteristics were simulated as a function of a parametric two- or one-dimensional grid of the IN node transition slope and the capacitive load of the OUT node. These simulations were arranged in tables that can be linearly interpolated or extrapolated by the tool.

7.4.1.2 Data-to-Data Timing Constraints

As stated above, data-to-data timing constraints are mandatory to generate valid dynamic non-monotonic logic networks. These constraints are used to apply a skew relationship between data pins. The definition of inter-data timing constraints is once again shown with the familiar example of a crossed $NAND_2$ cell. As illustrated in Fig. 7.11, a skew has to separate the arrival of two conflicting types to prevent competition of the Z node. This constraint is defined by the `timing_type` group of `non_seq_*` parameters of the Synopsys® LIBERTY library modeling standard format [15, 17]. To define the skew timing constraint between the rise (evaluation) of the *Type-B* IN_1 input and the fall (evaluation) of the *Type-A* IN_2 input, the `non_seq_setup_rising` timing constraint was defined relative to the IN_1 pin, and the IN_2 pin governed by the `fall_constraint` parameter values was denoted as $t_{(su,r-f)}$ in Fig. 7.11. These timing parameters were simulated and derived with the characterization tool based on appropriate criteria such as the OUT node voltage drop, excessive current drawn, etc.

7.4.2 Dynamic Library

The design characteristics described in pseudo-static *.lib* are somewhat different from those of dynamic *.lib* because of the distinction between the data evaluation and precharge phases which have completely different goals.

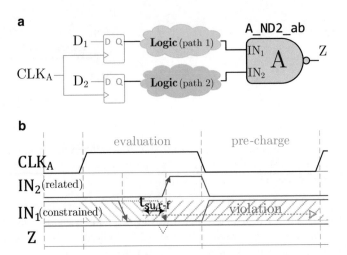

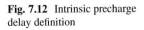

Fig. 7.11 Definition example of data-to-data timing constraint: (**a**) data-to-data timing constraint, data path example, and (**b**) data-to-data timing constraint, waveform example

Fig. 7.12 Intrinsic precharge delay definition

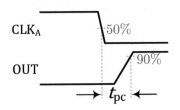

7.4.2.1   Design Metric Characteristics

For the evaluation phase period in the clock cycle, the characterization is identical to the pseudo-static *.lib*. By contrast, during the precharge phase there are a number of changes in the design metric assessments:

1. The intrinsic precharge delay is measured relative to the CLK edge and is irrelevant in terms of data propagation and thus only applies to the propagation of the precharged state, as illustrated in Fig. 7.12.
2. The data input capacitance C_{IN} is also irrelevant, since its discharge does not affect data propagation.
3. The CLK port input capacitance C_{CLK} is averaged to assess its switching:
$C_{CLK} = \frac{\int_0^T i_{CLK}(t)dt}{V_{DD}}$.
4. Power consumption: $P_{av} = P_{leak} + P_{dyn}$, where:

   a. $p_{dyn} = \frac{V_{DD}}{T} \int_0^T i_{V_{DD},dyn}(t)dt$, where all the current drawn is related to the transient short circuit caused by non-imminent propagation of the precharge state.
   b. $P_{leak} = V_{DD} \cdot I_{V_{DD},leak}$, where no stable short circuit condition is assumed.

7.4.2.2 Timing Constraints

The main goal of the dynamic DML library characterization is to identify its cells' timing constraints, which are defined to ensure that the tools have proper dynamic functionalities. One group of timing constraints has already been covered in the pseudo-static library subsection: the data-to-data constraints. Its characterization dynamic lib characterization is identical. This group aims to enforce the skew relationship between constrained data signals. The other group of constraints can be classified as data-to-clock but is only an abstraction, since DML clock pins are categorized as data pins. This group has several timing constraints that are reminiscent of the classic setup and involve constraints on sequential components. All of the following timing constraint parameters are characterized to be integrated into the Synopsys® LIBERTY library modeling standard format [15].

Setup/hold parameters—Timing parameters to avoid data signal transitions within the safety margins before/after edges of CLK_A, which are intended to isolate the evaluation and precharge phases and avoid data corruption.

7.4.3 Static Library

The standard CMOS-like library characterizes the design metrics of DML cells when the gate clock signals are disabled.

7.5 Benchmarks and Results

This subsection summarizes the results of the DML characterization methodology and its design flow.

7.5.1 Characterization

To evaluate the quality of the results on a *.lib* cell level, some fundamental design metrics of all the library cells were compared to their CMOS counterparts with the same technology node.

7.5.1.1 Performance

The evaluation delay measurements on the dynamic DML library were compared to the CMOS propagation delays on top of a two-dimensional grid of transition slope *vs.* capacitive load vectors. Figures 7.13, 7.14, and 7.15 speedup results consolidate

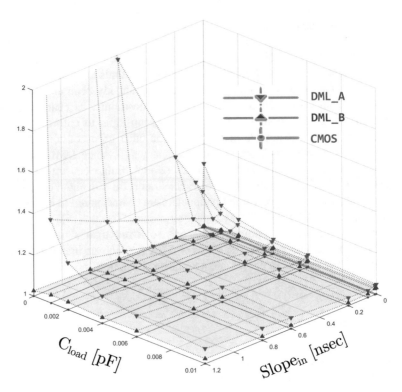

Fig. 7.13 Speedup of dynamic DML inverter *vs.* CMOS

the early theoretical assessment of DML's dynamic performance superiority. All the surface plots reveal a similar dynamic speedup pattern, where the most efficient type of DML cell (Figs. 7.13, 7.14) displays a significant performance boost, whereas the least efficient type of cell exhibits at worst the same speed as its CMOS counterpart. For example, a NOR_2 *Type-A* cell presents roughly a 25% performance gain for a nominal capacitance of 2fF and a rise time of 10 ps.

7.5.1.2 Area and Leakage

Area and leakage are usually tightly related, as shown in Fig. 7.16. It should be noted that the average leakage during the dynamic mode of DML exhibited similar behavior but was assessed somewhat differently because the precharge combination has more dominant weight over the rest, given its half cycle duration.

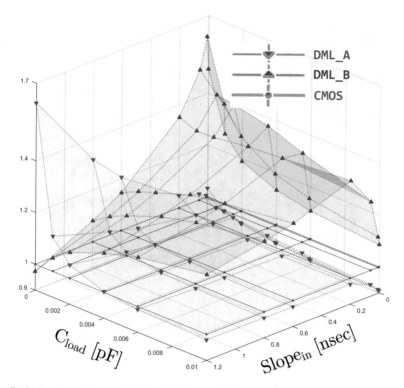

Fig. 7.14 Speedup of dynamic DML $NAND_2$ vs. CMOS

7.5.1.3 Equivalent Input Capacitance

Recall that the switching energy is linearly related to the equivalent capacitance $(E_{sw} \propto \alpha C_{eq} V_{DD}^2)$, which is dominated by the input DATA and CLK capacitances of the cells. Figure 7.17 shows that DML cells are more efficient in terms of data switching, but that continuous ($\alpha = 1$) clock toggling tips the scales in favor of CMOS.

7.5.2 Design Flow

The evaluation of the DML automated design flow results was based on the synthesis of a set of combinational RTL benchmarks. These benchmarks included typical logic blocks such as multiplexers, decoders, comparators, adders, complex Boolean logic functions with varying fan-ins, etc. Two concurrent synthesis processes were executed on the basis of the DML vs. CMOS equivalent standard libraries. To avoid black-box uncertainty regarding the commercial CMOS library, it was

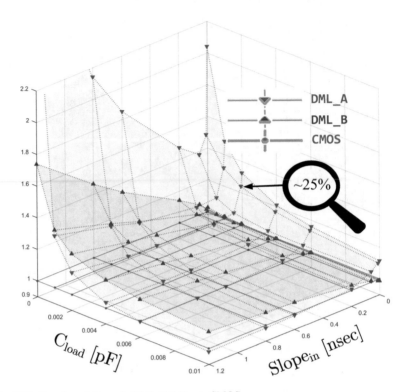

Fig. 7.15 Speedup of dynamic DML NOR_2 $vs.$ CMOS

independently laid out and re-characterized with similar conventions. Both libraries included identical minimal logic sets: $NAND_2$, NOR_2, and an inverter. The results fall into two groups: those that include and exclude non-unate logic. Tables 7.2 and 7.3 summarize the average design metrics of the unate and non-unate benchmarks for several gate count ranges.

Unate logic benchmarks exhibited better performing designs in terms of dynamic performance and slight power savings compared to the CMOS counterparts when operated statically. For example, the combinational barrel shifter presented in Table 7.4 dynamically sped up by 10%, with a power shift of about 20% in the static mode, whereas its gate count and area expanded by only about 4%. Furthermore, 50 generic combinational logic designs (with no particular functionality) were simulated and showed an average dynamic speedup of about 9%, a static power saving in the region of 20%, a similar gate count, and a slight area expansion of 8%.

By contrast, the non-unate benchmarks usually only had a dynamic performance gain, but lagged behind in terms of area and presented similar power consumption in the static mode (due to timing constraints which enforce delay insertion or logic duplication). For instance, the dual priority decoder presented in Table 7.5 had a similar static mode power consumption as CMOS due to a logic redundancy of about 21%, but still had a dynamic speedup of about 4%. Figure 7.18 depicts the E–

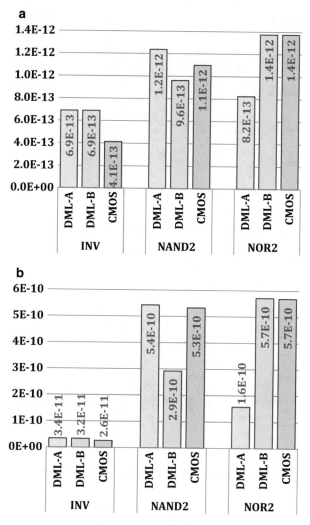

Fig. 7.16 Area and leakage comparison of cells: (**a**) area of cells m^2 and (**b**) *Static* DML leakage of cells W

D plane of the typical DML non-unate benchmarks relative to the static CMOS and graphically highlights the performance gains (linear x-axis) of the dynamic mode, although consuming much more power (logarithmic y-axis). The static mode exhibited much less extreme behavior and slight power savings at the expense of a moderate slowdown. The striking rise in power consumption of the dynamically operated DML was associated with a continuous refreshing of the entire logic networks at CLK speeds. Thus, the dynamic performance boost of DML non-unate designs should be carefully optimized to prevent persistent loads.

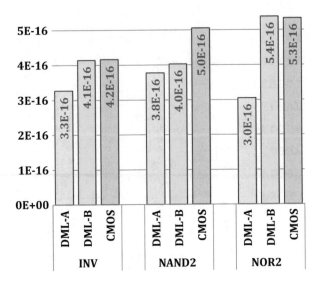

Fig. 7.17 Input capacitance F and switching energy of cells

Table 7.2 Average metrics for unate designs *vs.* CMOS, segmented by gate count ranges

Gate count range (sample space)	0–50 (20)	50–100 (15)	100–250 (10)	250–500 (5)
Area expansion	+12.5%	+6.9%	+5.8%	+3.1%
Gate count	+4.4%	+1.3%	+5.0%	−0.8%
Power shift (*static*)	−17.4%	−20.1%	−22.1%	−28.9%
Power shift (*dynamic*)	+315%	+323%	+299%	+340%
Speedup (*dynamic*)	+7.8%	+9.9%	+11.2%	+10.4%
Slowdown (*static*)	−32.7%	−31.1%	−37.7%	−41.5%

7.6   A Step Towards a DML Standard Flow: Conclusions

This chapter discussed ways to implement standard design tools and flows for
DML. It introduced a few ideas for DML characterization methodology for both
of its modes. Despite the numerous challenges of dynamic logic characterization
and design, the automated flow was able to generate timing compliant netlists and
in some cases exhibited better results in terms of design metrics as compared to
CMOS. The results here indicate that the DML design flow sometimes enables the
exploitation of DML advantages and provides a reasonably simple characterization
and design flow.

Table 7.3 Average metrics for non-unate designs *vs.* CMOS, segmented by gate count ranges

Gate count range (sample space)	0–50 (20)	50–100 (15)	100–250 (10)	250–500 (5)
Area expansion	+20.2%	+41.2%	+36.4%	+47.8%
Gate count	+4.3%	+11.8%	+15.2%	+21.8%
Power shift (*static*)	−3.20%	+3.40%	−0.70%	+5.20%
Power shift (*dynamic*)	+324%	+302%	+315%	+351%
Speedup (*dynamic*)	+7.4%	+8.4%	+16.1%	+12.2%
Slowdown (*static*)	−36.0%	−40.3%	−32.9%	−38.3%

Table 7.4 Design metric comparison *vs.* CMOS of unate designs

	Barrel shifter 16 bits	AND 128 bits	4to16 decoder	Avg. of 50 BMs
Area expansion	4.7%	1.9%	9.8%	8.5%
Gate count	3.4%	−2.3%	21.4%	3.1%
Power shift (*static*)	−20.4%	−24.7%	−6.0%	−20.3%
Power shift (*dynamic*)	243.0%	224.8%	305.7%	316%
Speedup (*dynamic*)	10.2%	9.4%	9.6%	9.4%
Slowdown (*static*)	−29.7%	−39.1%	−32.7%	−34.1%

However, as can clearly be seen from the comparison of DML standard flow designs to CMOS, the DML characteristics are far removed from the ones achieved by DML custom designs. This relates to static and dynamic power, delay, and area. For example, in all cases, DML custom designs showed an area reduction of up to 15%, compared to CMOS designs. In the next chapter, we show how the behavior of synthesized DML designs can be improved by introducing a DML-oriented synthesis. It is worth noting that by enabling voltage scaling on top of DML mode controllability, an extended E–D range can be achieved. The design of DML in conjunction with dynamic voltage and frequency scaling (DVFS) can be carried out in exactly the same manner as with standard CMOS. In other words, the DML library should be characterized for a desired number of voltages, and the physical implementation flow should be the same.

Table 7.5 Design metric comparison *vs.* CMOS of non-unate designs

	4-bit magnitude comparator	4-bit CLA adder	16 bit-dual priority decoder	Avg. of 50 BMs
Area expansion	49.9%	50.6%	37.9%	32.5%
Gate count	20.8%	15.1%	28.7%	10.5%
Power shift (*static*)	−7.2%	−11.4%	−0.9%	−3.6%
Power shift (*dynamic*)	301%	282%	327%	318%
Speedup (*dynamic*)	11.5%	12.0%	3.8%	9.9%
Slowdown (*static*)	−59.4%	−25.0%	−37.9%	−36.9%

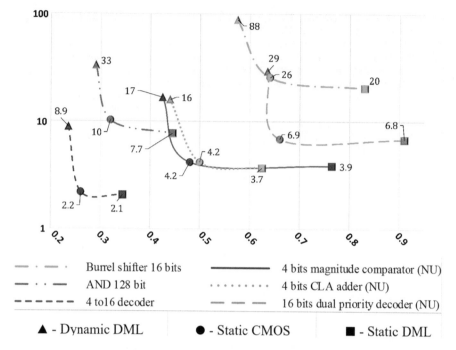

Fig. 7.18 E–D plane representation of typical non-unate benchmark designs. $E_{dyn}/cycle$ fJ $vs.$ $Delay$ ns

References

1. G. D. Hachtel, F. Somenzi, *Logic Synthesis and Verification Algorithms* (Springer Science & Business Media, 2006)
2. M.-B. Lin, *Introduction to VLSI Systems: A Logic, Circuit, and System Perspective* (CRC Press, 2011)
3. N. Weste, D. Harris, A. Banerjee, *CMOS VLSI Design: A Circuits and Systems Perspective*, vol. 11 (Addison-Wesle, Upper Saddle River, 2005), p. 739
4. A. Pal, A. Mukherjee, Synthesis of two-level dynamic cmos circuits, in *Proceedings of the IEEE Computer Society Workshop On VLSI'99* (IEEE, 1999), pp. 82–92
5. G. Yee, C. Sechen, Dynamic logic synthesis, in *Proceedings of the IEEE Custom Integrated Circuits Conference, 1997* (IEEE, 1997), pp. 345–348
6. M. Zhao, S.S. Sapatnekar, Technology mapping for domino logic, in *Proceedings of the 1998 IEEE/ACM International Conference on Computer-Aided Design* (ACM, 1998), pp. 248–251
7. T.J. Thorp, G.S. Yee, C.M. Sechen, Design and synthesis of dynamic circuits. IEEE Trans. Very Large Scale Integr. VLSI Syst. **11**(1), 141–149 (2003)
8. R. Hossain, *High Performance ASIC Design: Using Synthesizable Domino Logic in an ASIC Flow* (Cambridge University Press, Cambridge, 2008)
9. R. Krambeck, C.M. Lee, H.-F. Law, High-speed compact circuits with cmos. IEEE J. Solid State Circuits **17**(3), 614–619 (1982)
10. T. Williams, Dynamic logic: Clocked and asynchronous, in *Tutorial notes at the International Solid State Circuits Conference*, 1996

11. N.F. Goncalves, H. De Man, Nora: A racefree dynamic cmos technique for pipelined logic structures. IEEE J. Solid State Circuits **18**(3), 261–266 (1983)
12. G. Chartrand, *Introduction to Graph Theory* (Tata McGraw-Hill Education, 2006)
13. Y. Crama, P.L. Hammer, *Boolean Functions: Theory, Algorithms, and Applications* (Cambridge University Press, Cambridge, 2011)
14. J. Cortadella, A. Kondratyev, L. Lavagno, C. Sotiriou, Coping with the variability of combinational logic delays, in *Proceedings of the IEEE International Conference on Computer Design: VLSI in Computers and Processors, 2004. ICCD 2004* (IEEE, 2004), pp. 505–508
15. Liberty User Guides and Reference Manual, Synopsys, 2007
16. D. Harris, M.A. Horowitz, Skew-tolerant domino circuits. IEEE J. Solid State Circuits **32**(11), 1702–1711 (1997)
17. J. Bhasker, R. Chadha, *Static Timing Analysis for Nanometer Designs: A Practical Approach* (Springer Science & Business Media, 2009)

Chapter 8
Towards a DML Optimized Synthesis

In the previous chapter we discussed ways how to characterize DML cells into several libraries and use these with standard EDA tools. In this chapter we outline an optimized synthesis procedure for DML design. In a nutshell, this methodology involves changing certain steps within the *tools*. We present an algorithm for DML-optimized synthesis and show the implementation of this algorithm in Perl language. The synthesis results indicate that while this approach still has a significant room for improvement, it can boost performance gains and reduce the energy consumption of a design.

8.1 DML-Optimized Synthesis Challenges

Chapter 7 showed that the utilization of a new logic family within the standard design flow is extremely challenging. In practice, all the steps in the standard digital design flow (SDDF) assume that all the digital components are static CMOS or comply with the features of static CMOS. Given this assumption, highly automated EDA tools can place millions of gates on a single die and rapidly analyze their performance. As a result, any new logic family must come equipped with a solution for its integration with existing tools and design flows to enable smooth implementation within existing processes.

Logic synthesis, which is defined as the process of converting a high-level hardware description language (HDL) into a gate-level netlist (GTL), is one of the key components of the SDDF. Current synthesizers, which map behavioral register transfer level (RTL) code to specific standard cells, only support CMOS compliant gates that have static output levels driven by low-resistance devices, high-resistance capacitive inputs, and asynchronous combinational logic. Dynamic logic solutions like DM are characterized by different features and cannot be synthesized with current tools in a straightforward manner.

I. Levi, A. Fish, *Dual Mode Logic*, https://doi.org/10.1007/978-3-030-40786-5_8

Several methods have been put forward to integrate dynamic logic families into the SDDF [1–5]. Yee et al. [1] suggested a way to synthesize dynamic logic, in a way similar to procedure in [2] and [3], and described a CMOS-based solution with complex timing races that need to be addressed. Chappel et al. [4] introduced a system-level solution for integrating domino logic into the SDDF followed by a specific solution for synthesis, proposed by Parmar [5]. However, none of these approaches has considered logic such as DML that can function with several performance characteristics at the same operating corner. Similarly, they fail to take any of the constraints required for correct DML functionality into account.

This chapter details an approach to the synthesis of DML, as an initial fundamental step in the future full integration of DML into the SDDF. This approach manipulates standard synthesis tools and introduces a methodology to optimize the design which is attuned to the flexibility of DML and can cope with all the constraints related to DML. By applying this methodology, DML can reap the benefits of well-explored and highly optimized algorithms that are already part of the standard synthesis toolbox, but at the same time enable additional optimizations specifically for DML in a smooth integration process. To best illustrate this approach, more than 20 large benchmark circuits were synthesized in a 40 nm CMOS process and mapped to a DML library. The results demonstrate an improvement of up to 17% in timing and with up to 15% power reduction for large designs.

Specifically, this chapter describes:

- A fully automated design flow for DML enabling seamless integration into the SDDF, which make this new logic family applicable to modern large-scale designs
- A methodology for using DML to cut the delay of combinational logic in instances where this is required
- Results showing that this approach is effective in harnessing DML to reduce the power consumption of high-speed components without any additional changes to the implementation architecture, when performance can be sacrificed[1]

8.2   DML Synthesis Approach and Constraints

8.2.1   Constraints

In this chapter, only combinational logic is applied with the DML approach. In other words, the use of DML assumes a standard RTL design with the inference of CMOS registers that separate the paths of combinational logic. This approach

---

[1]Note that a designer can control and switch between the high-performance and low-energy modes on-the-fly.

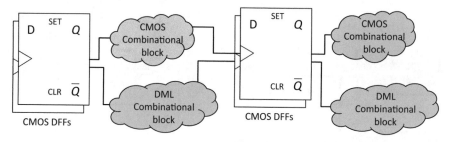

Fig. 8.1 Combinational block between two registers with DML and CMOS paths

enables the seamless integration of DML and static CMOS combinational blocks within a single system, as shown in Fig. 8.1, such that if the methodology does not generate an efficient solution for implementing a given path with DML gates, a design-time decision can be made to revert to static CMOS for a given path.

The synthesis of RTL to a standard CMOS library is a well-known procedure, and the constraints required to produce a logically equivalent netlist that meets the design requirements can be found in various commercially available EDA tools. The most common form of synthesis optimizes synchronous paths between two sequential elements (such as those illustrated in Fig. 8.1), ensuring that the max-delay (*setup*) and min-delay (*hold*) timing requirements are both met. A similar approach is used by DML, but by employing static CMOS sequential elements and requiring that the paths between these sequentials meet the standard setup and hold constraints. However, like dynamic logic, the efficient operation of DML in the *dynamic* mode relies on the precharge (PC) phase to drive the output to a predetermined level which at most should only change once during the evaluation phase, and thus requires additional constraints.

The next subsection defines three constraints that take these scenarios into account: namely, the correct precharge (CPC), footed gates (FG), and the single transition requirement (STR). The fundamentals behind these constraints have been already presented in the previous chapters of this book but are summarized here to make it easier for the reader to grasp the principles of the DML synthesis methodology. The definition of the constraints is followed by the synthesis methodology.

8.2.1.1 Correct Precharge (CPC)

In order to ensure that a DML gate starts the evaluation clock phase with a strong output level, the evaluation network must be cut off during the precharge (PC) phase. If the previous stages are all DML gates, their state during the PC phase is predetermined and known: *Type-A* gates precharge the output to "1" (V_{DD}) and *Type-B* gates predischarge the output to "0". The determination of the precharge input vector of a given gate leads to the selection of the gate type (*Type-A* or *Type-B*): The network that is cut off by this input vector is chosen to be the evaluation network, and the network that is enabled is chosen to be the precharge network.

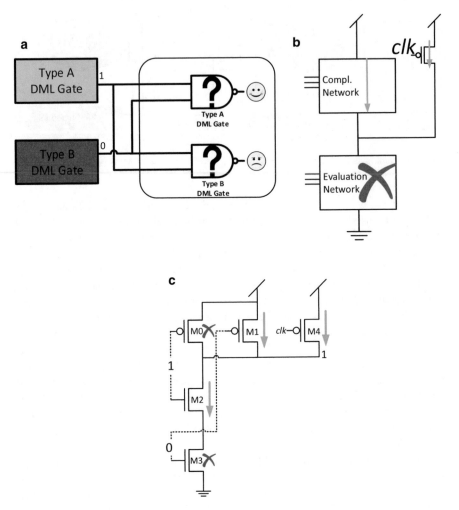

Fig. 8.2 CMOS-based DML gates in the PC phase. (**a**) NAND2 DML gate driven by an input vector of "10." (**b**) CMOS-based Type-A DML building blocks with correct precharge. (**c**) Type-A NAND gate with correct precharge

A simple example of a decision as to the type of DML gate is given in Fig. 8.2a, which shows a DML NAND gate driven by one *Type-A* and one *Type-B* DML gate. During the PC phase, the *Type-A* gate provides a "1" and the *Type-B* gate provides a "0" to the cascaded NAND gate. Given this precharge state ("10" or "01"), the structure of the NAND dictates that its PUN is conducting and its PDN is cut off. Therefore, we choose the PUN to be the precharge network and the PDN to be the evaluation network, which is the definition of a *Type-A* DML gate, as shown in Fig. 8.2b and c. A simple trick to figure out which gate should be used is that the gate type should match its precharge logic function, with *Type-A* matching a "1" and

Type-B matching a "0". For the NAND example above, a "10" input to a NAND is resolved to "1", matching a *Type-A* gate. An "11" input vector, on the other hand, would resolve to a "0", leading to the choice of a *Type-B* DML NAND gate.

8.2.1.2 Footed Gates (FG)

By complying with this constraint, a strong precharge level is ensured when cascading DML gates. However, one of the basic tenets of digital design with DML is that only combinational elements are dual-mode, while sequential elements are realized with standard CMOS gates. Unlike DML driven inputs, the state of CMOS-driven inputs is unknown during the PC phase, as is the state of the macro input ports. For this reason, a conducting evaluation network could occur, which can deplete the precharged output level before the evaluation starts.

The solution to this quandary is simply to use footed gates to implement the first DML stage after primary inputs (CMOS logic or input ports). This will ensure that the evaluation network is cut off during the precharge clock stage at the cost of an extra serially connected device.

8.2.1.3 Single Transition Requirement (STR)

According to the CPC constraints, the DML gate type will be chosen to match the determined input vector in the PC phase. This resolves standard cascading issues, but fails to take signal glitching into account. Due to the inherently unbalanced delays between the PDN and PUN of DML gates [6], there is a high likelihood of glitching on the internal combinatorial nodes prior to stabilization at their final state. If a node temporarily toggles to the wrong value and needs to be replenished, this will result not only in wasted power, but also in a considerable delay penalty if the node must be charged or discharged through the non-optimized network of its gate type. If the output of a DML gate switches more than once, this undermines the main advantages of using dynamic DML.

An example of this type of hazard is illustrated in Fig. 8.3. In this example, the first stage of NAND gates were chosen to be *Type-B*, and according to the FG constraint, footed gates were used. Thereafter, both the inverter and the output NAND were determined to be *Type-A* gates, according to the CPC constraint as described above. Now let's assume that the 3-bit input, in[2:0], transitions from 011 to 010, as illustrated in the associated waveforms in Fig. 8.3. The shorter path to internal node W will transition before the path to internal node Y, thus causing an unwanted temporary glitch to "0" on the output node Z.

In the above example, the selected configuration of DML gates (starting with the choice of *Type-B* NAND gates in the first stage) will lead to an STR violation, so that this configuration should be disqualified. However, most arbitrary circuits have at least one configuration that will not violate the STR constraint. The following procedure provides an algorithmic way to test adherence to this constraint and should be applied to every DML gate in the design.

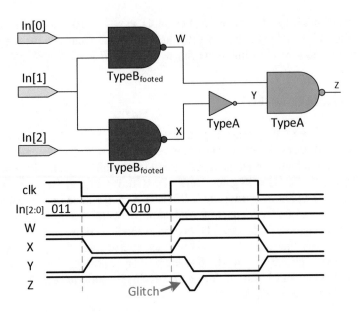

Fig. 8.3 Single transition hazard example

To conduct the STR check, the truth table of each logic gate is represented by a state diagram, with each input vector defining a state (i.e., 2^N states for a gate with a fan-in of N). Each state is assigned an output value, which is the gate output w.r.t the relative input vector, and the transitions between the states represent the transition of a single input. For example, the state diagram of a two-input NAND gate is shown in Fig. 8.4. To test the STR constraint, the initial state is the input vector due to precharge, and thereafter, the possible transitions should be followed to determine whether a particular order of arrivals of inputs would cause an output glitch.

The example in Fig. 8.3 illustrates an STR hazard. The next example shows a successful configuration. In this case, the first stage gates are implemented with *Type-A* footed NANDs, leading to the choice of a *Type-B* inverter and a *Type-A* output NAND, according to the CPC constraint, as depicted in Fig. 8.5a. The precharge input vector of the output NAND in this case is [in1, in2] =10, which is the initial state of the state diagram of Fig. 8.4. The only applicable transition that could possibly cause a glitch is the input arrival order *in1→ in2*, since the path to *in2* is longer than the path to *in1*. Therefore, the state machine is used to follow the state 10 through 00 (*in1* changes from "1" to "0") to 01 (*in2* changes from "0" to "1"). Continuing along the state diagram shows that the output of the NAND gate remains at "1" while traversing these states, thus indicating that the STR constraint is met. In this case, all three constraints (CPC, FG, and STR) are complied with, so that this configuration is valid for efficient DML implementation. Choosing a one *Type-A* and one *Type-B* gate for the first stage of NANDs would also have led to

Fig. 8.4 NAND2 single
transition test

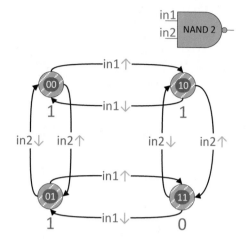

valid configurations. A methodology for implementing a design with DML logic
and validating these constraints is presented next.

8.2.2   The Approach

The methodology for DML synthesis takes all the requirements listed in the previous
subsection into consideration and converts the CMOS netlist into a DML netlist. The
gates in the netlist do not change; the only purpose of this process is to choose their
DML type: *Type-A* or *Type-B*, and whether the gate is footed or not.

The DML standard library cells in use should be fully characterized to include
all energy components and in particular the energy of the clocked elements.
Furthermore, a suitable setup and a hold delay characterization are needed to verify
both the timing and the DML constraints (CPC, FG, and STR).

The methodology consists of the following stages, as illustrated in Fig. 8.6:

1. **Determine the first-stage gates**. These gates have inputs that originate from
   primary inputs (sequential elements and macro inputs). They are implemented
   with footed *Type-A* or *Type-B* DML gates, according to the FG constraints. The
   first-stage gates can easily be found by tracing the outputs of the sequential
   elements and primary inputs of the design.
2. **Choose a DML type for each of the first-stage gates**. For N first-stage gates,
   there are 2^N possible solutions to choose their DML types.
3. **Derive the DML type of the internal gates by advancing through the netlist,
   stage by stage**. The DML type of each gate is determined by its logic function
   and the inputs at the PC phase to meet the CPC constraints.
4. **Calculate arrival times**. To ensure that all DML constraints are met, the timing
   for each node in the netlist needs to be calculated with standard static timing
   analysis (STA) methods. The capacitance and timing data are taken from DML
   timing libraries, characterized for the *dynamic* mode of operation.

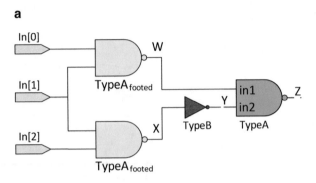

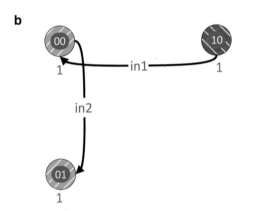

Fig. 8.5 Example of first-stage gate selection and single transition constraint checking. (a) Example logic block, including precharge values and selected gate types. (b) State transition diagram, showing glitch-free operation

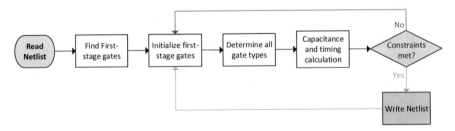

Fig. 8.6 Approach flowchart

5. **Check glitching requirements**. Every internal gate must be checked to ensure that its output cannot make more than one transition. This is done by advancing through the gate state diagram, as described for the STR constraint.

Fig. 8.7 Divide and
conquer—logic divided to
sub-netlists

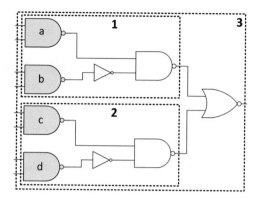

6. **If all constraints are met, a new GTL netlist is created that contains the DML gates.** Otherwise, the current configuration is discarded, and a different choice of first-stage gate types is made, by returning to **stage 3** of the flow.

Given the large number of possible first-stage combinations, a divide and conquer heuristic is used to eliminate redundant possibilities which cuts down significantly necessary runtime for finding a feasible solution. Large logic blocks are divided into smaller components and the constraints are applied to these sub-blocks. This is demonstrated in the example in Fig. 8.7, where a logic block is divided into three sub-netlists. The two inner sub-blocks (marked "1" and "2") are equivalent to the network in Fig. 8.5, which has only three possible solutions: AA, AB, and BA, as shown earlier. Therefore, all the other possible combinations can be omitted from the constraint checking of the higher-level module (marked "3" in Fig. 8.7). This reduces the number of possible input combinations for module "3" from 2^4 to 3^2 possibilities. The approach enables the application of this methodology to large designs.

8.3 Implementation Results

This subsection illustrates the application of the approach to several designs of varying sizes and complexities to test its performance when applied to real-life designs and challenging netlists. The flow was implemented in Perl.

8.3.1 Simulation Methodology

The simulation methodology is illustrated in the block diagram in Fig. 8.8. Several high-level RTL benchmarks of varying sizes and complexities were used for this evaluation, including some ISCAS'89 benchmarks. Each benchmark was synthesized using a Cadence RTL compiler (RC) and mapped to a 40 nm CMOS standard

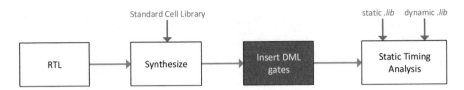

Fig. 8.8 Simulation flowchart

cell library to produce a CMOS gate-level (GTL) netlist. Each design was over-constrained to achieve the lowest possible delay (highest frequency) achievable with a standard CMOS library. The resulting GTL netlist was used as the input to the algorithm, which was applied to produce a DML GTL netlist. The resulting netlist was loaded back into RC and Synopsys PrimeTime (PT) to perform STA and analyze the results. The results for the two modes of DML operation were compared to the results of the standard CMOS implementation before the application of the algorithm.

As presented in the previous chapter, standard library characterization flows are not designed to correctly characterize DML gates. For this reason, a specialized characterization flow was developed in collaboration with Dolphin Integration based on the Dolphin Smash characterization tool to produce the Liberty timing files (.lib) that enable integration with standard EDA tools (the same library characterization as developed in the previous chapter). The DML library used here was developed in-house and characterized with this tool, subsequent to design, verification, and layout with Cadence Virtuoso. In particular, two separate libraries were constructed: a *static* mode library and a *dynamic* mode library as discussed in detail in the previous chapter. Proprietary setup and hold timing arcs were defined for each gate, and their characterization was extracted in addition to standard CMOS-like timing arcs. By using this approach, the results take into account all the features of the power and performance evaluation, including those attributed to clocking and precharge which do not occur in combinational CMOS gates. The outputs of this characterization process were verified by comparison to Spectre simulations on selected logic paths.

8.3.2   Simulation Results

To evaluate the methodology, each benchmark circuit was first synthesized by targeting the minimum delay (maximum frequency) with static CMOS libraries. The power, performance, and area results of the static CMOS implementation were extracted for reference comparison with the DML implementations. Then, the static CMOS GTL netlist was used as the input to the algorithm and mapped to DML libraries targeting the high performance *dynamic* mode. The resulting timing and power characteristics of 12 of the evaluated benchmarks are presented in Fig. 8.9.

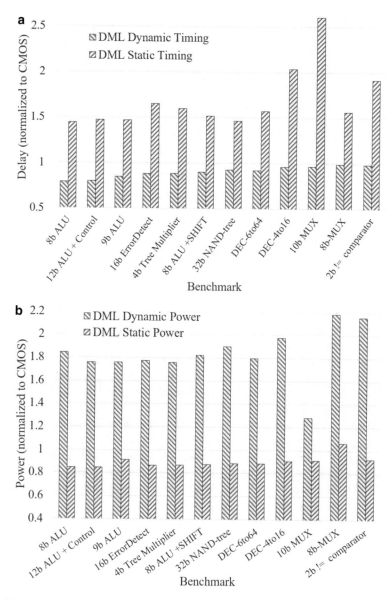

Fig. 8.9 Timing and power comparisons of the DML static and dynamic modes in 12 test-case benchmark circuits. (**a**) Delay (normalized to static CMOS). (**b**) Power (normalized to static CMOS)

The predicted superiority of *dynamic* DML over CMOS and the delay penalty for operating in *static* mode are clearly confirmed in Fig. 8.9a. Similarly, the power-saving advantages of operating in the *static* DML mode over CMOS and the power penalty for operating in the *dynamic* mode are shown in Fig. 8.9b. However, this

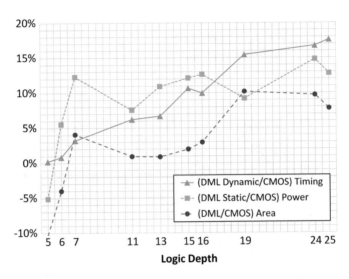

Fig. 8.10 Timing, power, and area improvement of DML compared to CMOS

figure points to the fact that the efficiency of using DML is strongly dependent on the characteristics of the underlying circuit. For example, using DML for the implementation of the 8b-MUX design was clearly a bad choice, since even in the static mode, this benchmark was inferior to CMOS. This suggests that an additional analysis could provide better insights into the relative effectiveness of using DML in a given design.

An extended analysis appears in Fig. 8.10, which plots the dependence of the achievable performance, area, and power improvements of DML as a function of the length of the logic path in which it is used. To extract these data, all of the logic paths in the implemented benchmarks were categorized as a function of the number of stages from startpoint to endpoint. The average delay, energy, and area of these paths was compared against their CMOS equivalents.[2]

These results clearly demonstrate that the speed improvement when operating DML in the *dynamic* mode monotonically increases with path depth, to reach as much as 17% for paths with 25 stages. This was expected, since the first stages of all DML paths are realized with footed gates, which are slower than standard DML gates. The effect of these gates is greater for shorter paths. When utilizing the *static* mode of DML, the energy efficiency increases sharply, with the power reduction exceeding 10% for paths with over 6 stages. This makes sense due to the inherent penalty of footed gates, which is considerable for short logic paths. The smaller footprint of DML gates also provides an area reduction for paths with more than 6 stages and stabilized at an average improvement of approximately 10% for

---

[2]The evaluation of both static and dynamic DML modes was applied to the same synthesized netlist, targeted in the *dynamic* mode.

paths with 19 or more stages. This also reflects the inherent efficiency of DML in constructing gates with smaller capacitance and superior performance.

Thus overall, DML was clearly shown to provide advantages over traditional CMOS in terms of the key features of digital design. By automating the integration of DML into the SDDF in conjunction with the design of smart controls to switch between the *static* and *dynamic* modes on-the-fly, DML can successfully meet all the traditional challenges of digital design.

8.4 Automated DML Synthesis: Conclusion

In this chapter, we overviewed a methodology for carrying out DML-compatible logic synthesis to efficiently map an RTL design onto a DML standard cell library. A fully automated DML synthesis flow compatible with standard tools and library characterization formats was presented and applied to numerous benchmark circuits. The synthesized circuits were evaluated for performance, power consumption, and area requirements in both the *static* and *dynamic* modes of operation, and exhibited average performance improvements as high as 17% for deep logic paths, with the ability to switch to an energy-efficient mode to save over 10% in energy. This was achieved with smaller silicon footprints, thus providing a 10% average area reduction. While we showed that integration of out-of-the-box logic families (such as DML) into the standard digital design flow is feasible, custom DML circuits still present better characteristics, leaving room for more improvements in the DML to SDF adaptation process. We hope that these results will pave the way for the usage of this logic family.

References

1. G. Yee, C. Sechen, Dynamic logic synthesis, in *Custom Integrated Circuits Conference, 1997, Proceedings of the IEEE 1997* (IEEE, Piscataway, 1997), pp. 345–348
2. A. Pal, A. Mukherjee, Synthesis of two-level dynamic CMOS circuits, in *IEEE Computer Society Workshop On VLSI'99. Proceedings* (IEEE, Piscataway, 1999), pp. 82–92
3. D. Samanta, A. Pal, N. Sinha, Synthesis of high performance low power dynamic CMOS circuits, in *Proceedings of the 2002 Asia and South Pacific Design Automation Conference* (IEEE Computer Society, Washington, 2002), p. 99
4. B. Chappell, P. Saxena, J. Vendrell, X. Wang, P. Patra, M. Venkateshmurthy, S. Jain, H. Krishnamurthy, S. Hussain, S. Gupta et al., A system-level solution to domino synthesis with 2 GHz application, in *2012 IEEE 30th International Conference on Computer Design (ICCD)* (IEEE Computer Society, Washington, 2002), pp. 164–164
5. D.M. Parmar, M. Sarma, D. Samanta, A novel approach to domino circuit synthesis, in *20th International Conference on VLSI Design, 2007. Held Jointly with 6th International Conference on Embedded Systems* (IEEE, Piscataway, 2007), pp. 401–406
6. I. Levi, A. Fish, Dual mode logic – design for energy efficiency and high performance. Access IEEE **1**, 258–265 (2013)

Chapter 9
Dual Mode Logic in FD-SOI Technology

Now that we have explored DML operation and efficiency in a conventional bulk CMOS, this chapter evaluates the DML technique in a relatively advanced 28 nm FD-SOI technology. Throughout, we provide fabricated ASIC measurements data to support the analysis and theoretical foundations presented in the previous chapters. In addition we show how DML logic can utilize the unique features of an ultra-thin body and box (UTBB) fully depleted silicon on insulator (FD-SOI) technology to achieve high-speed and energy-efficient designs for a wide range of supply voltage operations. This chapter starts with a brief comparison of DML and conventional static and dynamic CMOS logics for NAND–NOR chains in 28 nm FD-SOI. This basic analysis is followed by the construction of a real-life benchmark, a two-stage pipelined multiply-accumulate (MAC) circuit which was selected to assess the advantages of DML in terms of speed, energy, and area as compared to a conventional CMOS design. We show that the self-adjusted DML MAC achieves both a performance boost of up to 92% with 16% less energy consumption than the equivalent standard CMOS implementation. The energy saved can reach up to 35% when the low-power (fully static) mode is enabled. In addition, the DML MAC occupies 25% less area.

9.1 UTBB FD-SOI Technology

Ultra-thin body and box (UTBB) fully depleted silicon on insulator (FD-SOI) is an emerging technology that draws on the conventional planar bulk CMOS process while keeping pace with the efficiency improvements projected by Moore's law [1, 2]. Because it constitutes a relatively simple evolution from the conventional CMOS process, UTBB FD-SOI can provide reduced die size and power consumption along with increased performance and functionality. These benefits are achieved without radical complex manufacturing steps. ST 28 nm UTBB FD-SOI technology devices are planar CMOS transistors fabricated in a 7 nm layer of silicon placed over a 25

© Springer Nature Switzerland AG 2021
I. Levi, A. Fish, *Dual Mode Logic*, https://doi.org/10.1007/978-3-030-40786-5_9

nm buried oxide (BOX). The ultra-thin silicon film ensures that all the electrical paths between the source and the drain are confined to the gate region, which yields improved subthreshold slope and drain-induced barrier lowering (DIBL) [1]. The fully depleted channel of devices avoids the issue of random dopant fluctuation, which lessens device variability. Since the BOX constitutes a dielectric isolator of the source and drain from the underlying n/p-well, which eliminates the parasitic effects of body biasing, the feasible body bias range is larger than standard bulk CMOS technologies and makes it one of the main calling cards of UTBB FD-SOI technology. It authorizes high performance and low power for increased energy efficiency [3].

In this chapter, the DML technique is evaluated in 28 nm STM UTBB FD-SOI technology to demonstrate high-energy-efficient designs in a wide supply voltage operation range. The dependence of power dissipation (and performance) on the supply voltage dictated the circuit design for dynamic voltage and frequency scaling capability [4–6]. As shown below, the combination of DML properties and the extended body bias capability of the UTBB SOI technology makes it possible to design highly energy-efficient digital systems.

9.2 DML Design Optimization in UTBB FD-SOI Technology

This subsection focuses on design optimization and the related properties of the DML primitives in UTBB FD-SOI technology. The energy, delay, and sensitivity of the DML technique to process variations is compared to conventional static and dynamic CMOS logic designs.

9.2.1 Design Optimization

Figure 9.1 illustrates the design strategy for (a) Type-A un-footed, (b) Type-B unfooted, (c) Type-A footed, and (d) Type-B headed DML gates, respectively. In UTBB FD-SOI there are two "well"-topologies, which are known as the conventional well (CW) for regular voltage threshold (RVT) devices and the flip-well (FW) for low-voltage threshold (LVT) transistors. When implementing the CW, the reverse body bias (RBB) range can be as low as -3 V and for FW, the forward body bias (FBB) as high as 3 V [5]. As depicted in Fig. 9.1(a–d), we assumed that the FW configuration fully exploited the powerful knob (not available in nanoscale bulk CMOS technologies) provided by the FBB to compensate for variations and/or boost performance in a broad power supply operating range [7]. In what follows, we only consider symmetrical back biasing with $V_{bb} = GND$ in the static operation mode and three possibilities in the dynamic operation mode ($V_{bb} = GND/VDD/2 * VDD$). The physical design guarantees that for all

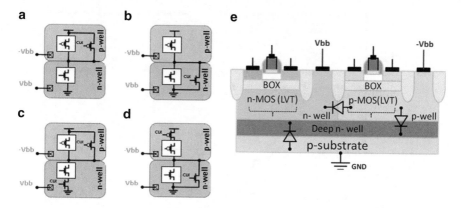

Fig. 9.1 LVT transistors for DML gate design in UTBB FD-SOI technology: (**a**) Type-A un-footed, (**b**) Type-B un-footed, (**c**) Type-A footed, (**d**) Type-B headed, and (**e**) cross-section

possible operating modes, all the parasitic diodes associated with the devices are maintained in the reverse mode (see Fig. 9.1e).

9.2.2   Performance and Robustness Analysis

The DML approach was compared to its CMOS static and dynamic counterparts over a large power supply range (V_{DD}: 0.3–1 V). For the first benchmark, we used a test chain composed of 20 fan-outs of four (FO4) interleaved NAND–NOR gates (one footed/headed gate for every five gates to reasonably confine the short circuit energy [8]). The NOR gates were implemented in the efficient A DML type and the NANDs were implemented in the efficient B DML type, thus resulting in a structure similar to a np-CMOS/NORA design. To ensure a fair comparison, the same choice was made for the CMOS dynamic gates. To prevent charge loss and improve noise tolerance, a weak keeper transistor was used for the CMOS dynamic gates.

The static CMOS gates were sized for a symmetrical switching delay at a power supply voltage of 0.6 V, and the pull-down network strength was set to be the equivalent of a single NMOS sized with $W = 240$ nm. The strength of the evaluation network of both the DML and CMOS dynamic gates was set to be equivalent to the corresponding transistor network of the static CMOS gates. In contrast, for the DML design, the complementary network transistors were sized with $W = 120$ nm to reduce intrinsic capacitances. Thus, in the dynamic operation mode, the fast transition was used for evaluation and the slow transition was employed for precharge. Bear in mind that fast evaluation is mandatory, since circuit performance is determined by the length of the total critical path under evaluation. However, while in precharge, a slower transition is possible since the precharge is a parallel fast process. Clearly, this favors speed in the dynamic operation mode at the price of

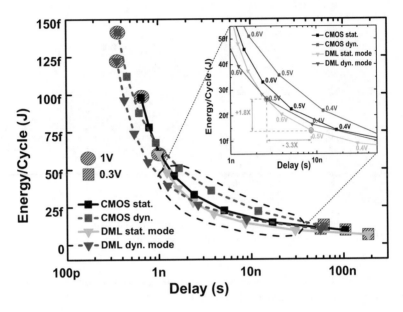

Fig. 9.2 Comparison in terms of the energy delay of DML and CMOS (static and dynamic gates)

an increased delay in the static mode. Finally, the precharge/discharge transistors (of both DML and conventional dynamic logic gates) were sized for the same strength as a NMOS sized with $W = 120$ nm.

Figure 9.2 shows the comparative energy–delay results obtained from the simulation setup presented above. Because of the asymmetric transistor sizing, when in the static mode, the DML design exhibited lower energy consumption (-37% on average) than its static CMOS counterpart, but experienced an increased delay ($+64\%$ on average). After switching to the dynamic mode, the DML design achieved a $3.2\times$ frequency boost on average, with an energy consumption increase of about $1.9\times$, on average. In the dynamic mode, the DML chain was faster (20% on average) than the np-CMOS design but also consumed less energy (20% energy consumption saved on average) by avoiding the keeper transistor (and the additional driving inverter).

One of the interesting features of the FD-SOI technology is its extended back-biasing capability. This FD-SOI feature enables the energy–delay space extension of the DML design. This is depicted in Fig. 9.3, where three different V_{bb} voltage levels are plotted in the dynamic operation mode. The strong body effect ($\sim$60–80 mV/V) of the technology effectively trades off energy for performance at a given V_{DD}. By contrast, a combination of dynamic body bias and voltage scaling can be exploited for energy savings. For instance, a comparable maximum operating frequency as the one achieved in the static mode by the DML chain for $V_{DD} = 0.9$ V can be produced in the dynamic mode by operating at $V_{DD} = 0.5$ with $V_{bb} = 1$ V. Thus, $\sim$40% energy can be economized.

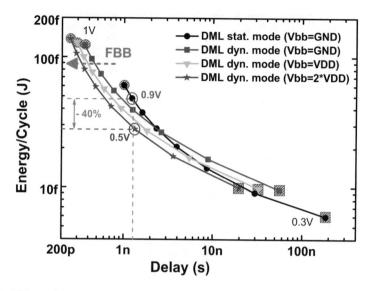

Fig. 9.3 Effects of the extended forward body bias voltage on the DML NAND–NOR chain in 28 nm UTBB FD-SOI

To analyze the effects of process variations on the delays produced by each circuit, Monte-Carlo (MC) simulations on $10K$ runs were conducted. The results appear in Fig. 9.4, which shows the delay variability (defined as σ/μ) and 3-sigma delay (defined as $\mu + 3\sigma$). Analysis of the dynamic CMOS logic gates indicated that their variability was significantly higher (in particular at lower V_{DD}) than that of the static CMOS logic. This is most likely due to the positive feedback associated with the keeper transistor which not only degrades the speed of dynamic logic (because of the current contention with the evaluation network) but also increases the delay variability [9].

As noted above, this positive feedback is not needed for DML gates and has a favorable impact on delay variability in the dynamic operation mode. Figure 9.4a shows that the delay variability of the DML design (in both the static and dynamic operating modes) was always under 7% for the whole V_{DD} range tested. The MC results in Fig. 9.4b show that as expected, the DML chain operating in the static mode had the largest 3-sigma delay. By contrast, the DML design presented the lowest 3-sigma delay when operating in the dynamic mode. Bear in mind that the 3-sigma delay of the dynamic CMOS chain degrades rapidly for lower-power supplies (a consequence of the larger variability) and becomes about $2\times$ that of the dynamic DML design for $V_{DD} = 0.3$ V.

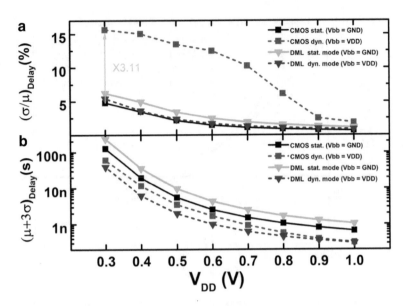

Fig. 9.4 Variability of the designs from $V_{DD} = 0.3$ to 1 V

9.3   Design: Dynamically Adaptable Multiply-Accumulate Circuit in 28 nm FD-SOI

This subsection discusses the potential of the DML technique to provide E–D optimizations at runtime by implementing of an 8×8 bit dynamically adaptable multiply-accumulate (MAC) unit implemented in a 28 nm FD-SOI process. This is a meaningful benchmark since the MAC block is the prime circuit for assessing the operating characteristics in many digital signal processing (DSP) applications such as convolution, digital filtering, and the fast Fourier transform (FFT) [10, 11]. When performance is the goal, the mixed operating mode is enabled and a low-complexity self-adjustment mechanism is implemented at runtime to identify the gates/blocks that need to work in the energy-efficient mode (i.e., static mode), while those that are part of the longer logic (critical) paths can operate in the faster dynamic mode. In addition, the MAC circuit can run in the fully static mode to ensure the lowest energy consumption when speed is secondary.

Figure 9.5 illustrates the top-level architecture of the two-stage pipelined MAC unit, where low-complexity extra clock control circuitry blocks are required for the DML design. The first stage originates from the column bypassing the partial product reduction tree (CB-PPRT) [12] which was custom re-designed for the DML-based architecture. The CB-PPRT provides reduced partial products in the carry-save (CS) format that are added to the result (in the conventional binary format) generated by the second stage in the previous clock cycle. This function is handled by the 3:2 DML compressor (using simple DML full adders (FAs) [13]),

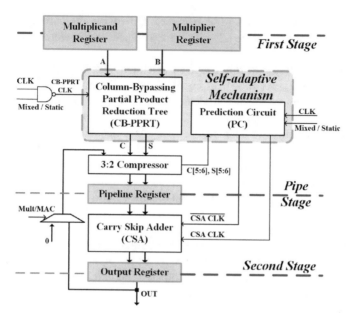

Fig. 9.5   Architecture of the DML MAC

which generates a CS-format number for the second stage. In the second stage, the DML carry-skip adder (CSA) generates the final output.

The DML MAC incorporates a self-adjustment mechanism triggered in mixed-mode operation. This mechanism operates in the first stage and dynamically controls the operation mode (static or dynamic) of gates/blocks in the DML design. Part of the self-adjustment mechanism is embedded in each cell of the CB-PPRT, and the rest is implemented in the prediction circuit (PC); both are explained in more detail in Sects. 9.3.1 and 9.3.2. The architecture supports two arithmetic operations: MAC and multiply.

9.3.1   DML Column Bypassing Partial Product Reduction Tree

The CB technique suggested by Wen et al. [12] saves dynamic energy consumption by cleverly exploiting the observation that the outputs of all the FAs in column j of the PPRT can be predicted early if the corresponding bit in the multiplicand Aj is at the low logical level. Thus, depending on the number of zeros in the word of the multiplier A, the switching activity of a certain number of FAs can be reduced while maintaining their correct functionality.

For this purpose, the generic FA in the PPRT is modified to add two tristate gates at the input operands and a 2:1 multiplexer at the output sum controlled by the corresponding Aj bit [12]. When Aj is 0, the inputs of the FA are disabled and the multiplexer allows the FA to be bypassed. Otherwise, the normal operation is executed. More details on the CB-PPRT are available in the works by Wen et al. [12, 14].

The design presented in this chapter implements the Wen et al. technique in a different way [12]. As shown in Fig. 9.6a, when the mixed-mode operation is enabled, FAs belonging to column j which can be bypassed ($Aj = 0$) are selected for pseudo-static operation, while the remainder run in the dynamic mode. This permits energy savings without incurring delay penalties. Further energy reductions are guaranteed by the bypass circuitry embedded in the modified DML mirror FA, as depicted in Fig. 9.6b and c. When Aj is 0, the extra bypass transistors (M0–M6) avoid energy-hungry glitches while setting the correct outputs of the DML FA.

Figure 9.6d depicts the behavior of the modified DML FA when running in the static and mixed modes. When launched in the DML MAC static mode, the bypass circuitry reduces switching activity, as discussed in Wen et al. [12]. In the DML MAC mixed mode, with $Aj = bp = 0$, the output carry (Co) is set early during the precharge phase and is maintained at 0 by the incoming inputs in the evaluation phase. The correct output sum (So) is computed (no glitches can be generated) by the transistors M3–M6 in the evaluation phase.

9.3.2 Adaptive Final DML Carry-Skip Adder

As extensively discussed, the probability that the longest carry path of an adder circuit will be triggered by the input operands is very low, since in most cases, the actual carry path propagation in the addition operation is much shorter than the adder's low-probable critical path [15]. Concretely, this means that the final DML CSA can complete most operations at a slower speed (static mode), thus conserving precious energy and boosting its performance (dynamic mode) when a low-probable set of input vectors triggers longer carry propagation paths. This capability to adjust the computation to the incoming input vectors is provided by the PC, which enables even more energy savings when the MAC is operating in the mixed mode.

As depicted in Fig. 9.7a, when the DML MAC operates in the mixed mode, just by checking the signals $S[6 : 5]$ and $C[6 : 5]$ produced by the CB-PPRT, the PC can detect the existence of a longer carry path propagation than the authorized timing budget and switches the operation of the final DML CSA to the dynamic mode. Crucially, the signals $S[6 : 5]$ and $C[6 : 5]$ are processed in the first stage of the MAC to produce clock signals for the DML CSA in the second stage without impacting the delay of the final adder. In addition, these signals are not on the critical path of the MAC circuit in the first pipeline stage, so that the energy savings are achieved without compromising on performance.

Figure 9.7b and c illustrates the two operating modes of the adaptable 16-bit DML CSA. To avoid timing violations and save energy when the DML MAC is operating in the mixed mode, the final adder is optimized to achieve a similar delay in both the static and dynamic operation modes. Figure 9.7d provides a behavioral

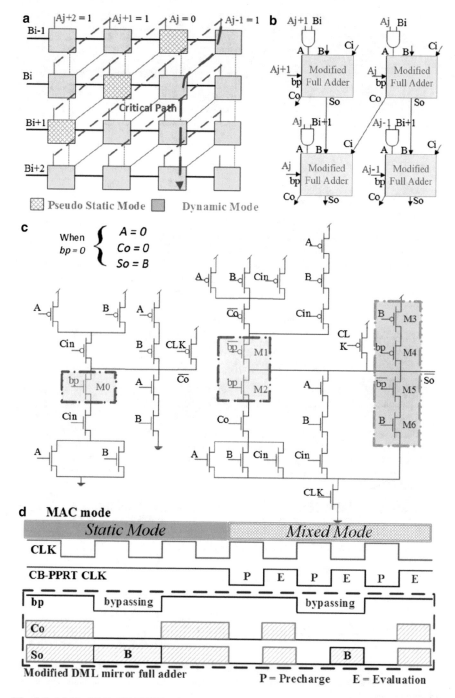

Fig. 9.6 (**a**) The DML CB-PPRT in the mixed-mode operation. (**b**) Sketch of a group of modified FA cells in the CB-PPRT. (**c**) Schematic of the modified DML mirror FA. (**d**) Behavioral waveforms of the DML FA in the static and mixed modes

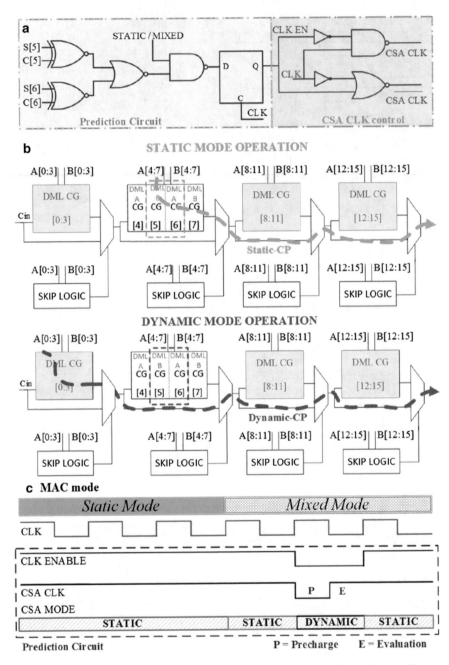

Fig. 9.7 (**a**) Schematics of the prediction circuit, the critical path of the DML CSA operating in the (**b**) static mode and (**c**) dynamic mode and (**d**) behavioral waveforms in the static and mixed operation modes

depiction of the working modes of the DML adder when the static and mixed MAC operation modes are enabled. As shown, the CSA only operates in the dynamic mode when the active low CLK_EN signal has been asserted (within the mixed-mode operation of the entire MAC).

9.3.3 Measurement Results

The self-adjusted MAC architecture was fabricated in a 28 nm FD-SOI technology. A conventional CMOS counterpart, based on the same two-stage pipelined architecture as our DML MAC (in other words, a CB-PPRT + 3:2 compressor for the first pipelining stage and a CSA for the second stage), was also fabricated to act as a reference design for comparison purposes. The CMOS MAC was designed to exploit a full custom approach, which is roughly 50% less energy greedy than the equivalent standard cell design while also improving the frequency and area occupation.

Below, we present the measurement results and examine the circuit in terms of energy, performance, robustness, and area. These results are based on measurements of ten test chips bonded to a QFN64 package. The die photograph is shown in Fig. 9.8a. To allow for modular validation, the experimental framework illustrated in Fig. 9.8b included a printed circuit board that was interfaced to a Virtex-5 field-programmable gate array (FPGA) development board used to stimulate and control the device under test through an FPGA mezzanine card connection. Local (and regulated) power supply generators supplied power to the isolated power domains of the test circuits on the die.

A controllable temperature isolation circuit monitored the device, allowing for external-temperature analysis in the 0–70 °C range. Figure 9.8c shows the physical implementation of the 8 × 8 CMOS and DML MACs. Despite the extra circuitry needed to implement the self-adjustment mechanism, the DML MAC occupies less silicon area because of its unconventional sizing strategy described in detail by the authors in Chaps. 2 and 3 (and in [16, 17]), which uses minimum-sized transistors for most of the DML gates. As a result, the DML design achieves a 25% reduction in total area as compared to the CMOS implementation (2084 versus 2811 μm^2). The extra sub-circuits needed for the self-adaptive mechanism (i.e., PC and CLK controls) cause very small area overhead (i.e., roughly 3% of the total area) in the DML MAC design.

9.3.3.1 Energy Consumption and Performance

For a fair comparison, both DML and CMOS designs were optimized for a supply voltage of 0.6 V, which is consistent with the contemporary trend of energy-efficient designs in advanced technology nodes [18]. Energy per operation (E/Op) was evaluated for different activity factors (i.e., $\alpha = 0.1, 0.3$, and 0.5); for each activity

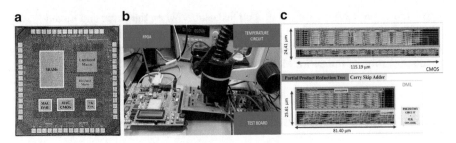

Fig. 9.8 (**a**) Micrograph of the test chip, (**b**) experimental setup, and (**c**) layouts of the CMOS (top) and DML (bottom) designs

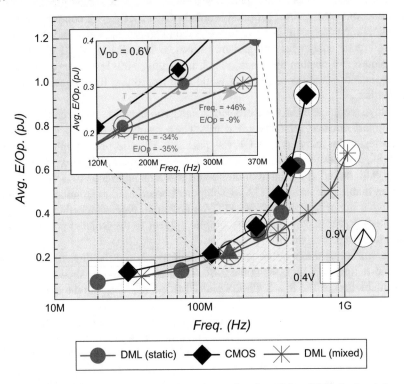

Fig. 9.9 Average E/Op versus frequency for V_{DD} ranging from 0.4 to 0.9 V; the inset shows a comparison at $V_{DD} = 0.6$ V ($\alpha = 0.3$)

factor, the E/Op was averaged over a set of 1k input transitions, which corresponded to this type of activity. For the DML MAC, energy measurements included all the additional control circuits. Figure 9.9 depicts the energy frequency characteristics of the circuits for V_{DD} ranging from 0.4 to 0.9 V with $\alpha = 0.3$. Our circuit, which operates in the mixed DML mode, exhibited lower energy consumption and higher frequency than the CMOS design for the entire power supply range; it had a maximum speed advantage of 92% at 0.9 V (55% on average) with average energy

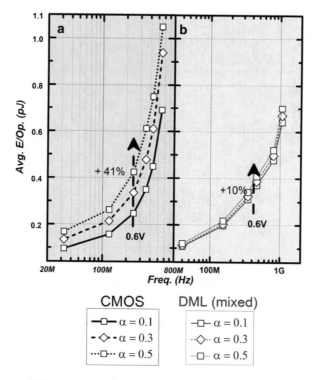

Fig. 9.10 Average E/Op versus frequency for (**a**) CMOS and (**b**) DML mixed design evaluated for different activity factors α from 0.1 to 0.5 (T = room temperature)

savings of $\sim$16%. The zoomed-in subfigure compares the CMOS and DML circuits at the device-sizing optimization point of 0.6 V; the MAC circuit operating in the DML static mode yielded an energy reduction of 35% at the expense of a 34% frequency reduction. By contrast, when in the mixed DML mode, it outperformed the CMOS in terms of speed by 46% and reduced the E/Op by 9%.

As shown in Mahant-Shetti et al. [19], energy savings are achieved by avoiding/reducing glitches in the PPRT of the multipliers. Thanks to the inherent properties of dynamic logic-based styles, the DML MAC design is glitch-free when operating in the mixed mode. Furthermore, as a result of its adaptive mechanism, low-activity nodes in the design rarely operate in the DML dynamic mode (i.e., when the DML MAC is in the mixed mode). These two features lead to the energy savings of the DML MAC in the mixed operation mode as compared to the static CMOS approach. To illustrate this behavior, energy and frequency measurements were obtained by asserting inputs to the MACs that corresponded to different activity factors. As shown in Fig. 9.10, at the highest activity factor, whereas the static CMOS approach increased its energy consumption by as much as 41% (38% for DML MAC in static mode), our circuit in the mixed DML mode increased the energy by only 10%.

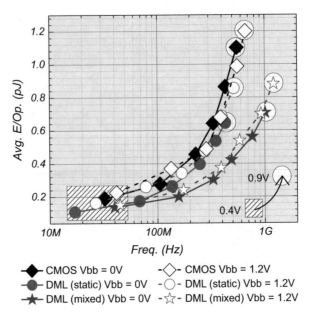

Fig. 9.11 Average E/Op versus frequency under symmetric forward body biased voltage ($\alpha = 0.5$, $T =$ room temperature)

One of the main features of the FD-SOI technology is its wide allowable FBB voltage range for LVT transistors [7, 20]. The efficiency of this feature, i.e., symmetric FBB of $V_{bb} = 1.2$ V, was evaluated and is depicted in Fig. 9.11. Both the DML and CMOS designs increased their operating frequency by roughly 20%. However, as expected, this came at a price, in that the E/Op also increased by 33%/20% for DML MAC in the mixed/static modes and by 17% for the CMOS. In the DML design, the FBB can be used to boost the frequency of the MAC in the static mode so that it is practically equal to the CMOS frequency while at the same time reducing the energy consumption by 14%.

9.3.3.2   Robustness and Process/Voltage/Temperature Variations

The impact of process variations with respect to both energy and frequency was analyzed over all the dies. The means and standard deviations of both the frequency and energy were calculated (denoted by μ_{FREQ}, σ_{FREQ}, μ_{EOP}, and σ_{EOP}). The values appear in Fig. 9.12 for $V_{DD} = 0.6$ V. The DML MAC in the mixed mode of operation exhibited lower variability (in terms of σ/μ). The mean frequency of the DML MAC in the mixed mode confirmed the results reported in Sect. 9.3.3.1 (46% faster). In terms of variability (σ/μ), the MAC results in the static DML mode demonstrated its greater robustness (in terms of both energy and frequency) as compared to the CMOS circuit.

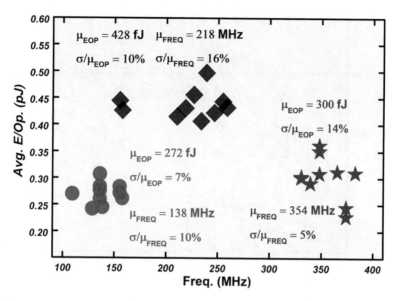

Fig. 9.12 Variability of E/Op versus frequency over 10 test chips ($V_{DD} = 0.6$ V, $\alpha = 0.5$ and $T =$ room temperature)

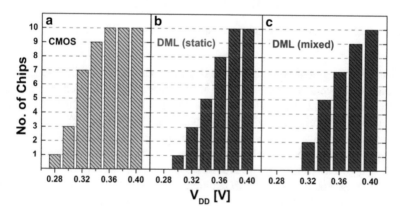

Fig. 9.13 Minimum V_{DD} (measured) over 10 test chips of (**a**) conventional CMOS, (**b**) DML static, and (**c**) DML mixed designs

To further probe the functionality of our design in different DML modes, the minimum operating voltage was evaluated (over all the dies). Figure 9.13 shows the cumulative distribution of the number of dies that were functional per supply voltage at the sub- to low-near-threshold voltage regions. The results in Fig. 9.13 demonstrate that all the measured DML MAC circuits were fully functional at 400 mV in all modes, and that in the static mode, even a lower voltage of 380 mV was achieved. These results are very similar to the CMOS circuits, where all the samples were operational at 360 mV.

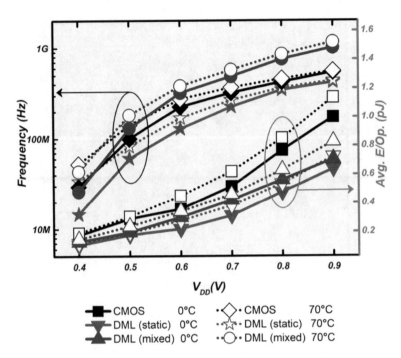

Fig. 9.14 Effect of temperature in frequency and average energy per operation

One of the key goals was to explore the sensitivity of the DML-based MAC to temperature, especially in the near-threshold domain. Figure 9.14 charts the energy and frequency values against temperature for low 0 °C and high 70 °C (device-external temperature). The y-axis for frequency appears on the left and the energy y-axis appears on the right of Fig. 9.14. The comparison indicates that in cases of V_{DD} exceeding 0.5 V, the DML MAC in the mixed mode outperformed the CMOS design and showed less sensitivity throughout the entire power supply range. However, as expected, both the CMOS circuit and the DML MAC in the static mode were significantly affected in terms of frequency at low voltages. In the static mode, the DML MAC was more sensitive in terms of frequency, but this only became substantial for supply voltages below 450 mV. Nevertheless, the CMOS and the DML MAC in the static mode only exhibited a slight increase in terms of energy in the near-subthreshold region.

9.3.3.3 Comparison to the State of the Art

The self-adjusted DML MAC is compared to several recent descriptions of MAC/multiplier circuits in Table 9.1. This comparison served to identify implementations with similar technology nodes and architectures that also cover a similar voltage supply range. Table 9.1 lists two low-voltage (400 mV) designs: an

Table 9.1 Comparison to state-of-the-art MACs

	A-SSCC '14 [22]	MIXDES '16 [21]	DML (in this chapter)						TVLSI '17 [23]
Technology design (N. Bits)	28 nm FD-SOI 16 × 16 MAC	28 nm FD-SOI 8 × 8 MAC	28 nm FD-SOI 8 × 8 MAC						65 nm 16 × 16 Multiplier
			Pipelined column bypassing						
			DML		CMOS	DML		CMOS	
Architecture	Deep pipelined modified Baugh-Wooley	Pipelined MAC	Static	Mixed		Static	Mixed		1-stage multiplier
V_{DD} [V]	0.4	0.4	0.4			0.8			0.8
Latency [Cycles]	32	2	2			2			1
Frequency [MHz]	8.75[a,b]	51[b]	**20.3** (10.15[c])	**40.1** (20.1[c])	32.5 (16.3[c])	**367.5** (183.75[c])	**800.4** (400.2[c])	428.8 (214.4[c])	370[b]
E/Op. [pJ]	0.39[b]	0.09[b]	**0.08** (0.32[c])	**0.11** (0.44[c])	0.13 (0.52[c])	**0.39** (1.56[c])	**0.49** (1.96[c])	0.6 (2.4[c])	3.8[b]
Area [μm²]	7569	2208	**2084** (**6271[d]**)		2811 (7029[d])	**2084** (**6271[d]**)		2811 (7029[d])	–
Test chip	✓	–	✓						–

The bold represent the added value of this work

[a] Frequency is normalized to a two-stage pipelined MAC

[b] The data are extrapolated from related papers

[c] Energy and frequency were evaluated for a 16-bit design according to the following formula in [21] and [22], respectively: $E/op_{(8\text{-bit})} * 4 = E/op_{(16\text{-bit})}$; $Freq_{(8\text{-bit})} * 0.5 = Freq_{(16\text{-bit})}$

equivalent 8 × 8 MAC presented by Vatanjou et al. [21] and the deeply pipelined (32 stages) 16 × 16 MAC put forward by Reyserhove et al. [22]. Cerqueira and Seok [23] reported a comparison point at a higher-voltage regime (800 mV) with a one-stage multiplier architecture [23]. Note that for the sake of comparison, we also provide a projection of the measurement values of the DML MAC to a 16 × 16 architecture by multiplying the area utilization of the PPRT by four and the final adder by two, the energy by four, and dividing the frequency by two [22, 23]. This is reasonable since the energy/delay advantages of the proposed implementation are likely to be maintained (if not better) when the size of the architecture scales for larger operands.

At 400 mV, the DML MAC in the static mode achieved $\approx \times 1.2$ in performance while reducing the energy consumption by ≈18% as compared to [22]. In the mixed mode, performance increased by more than twofold, and the energy was only slightly degraded by 5%. When evaluated against [21], in the mixed mode, the DML MAC results were similar; however, the static mode provided an 11% energy reduction and a ≈6% area reduction (bear in mind that [21] exploited asymmetric body biasing that can also be applied to our design; the results are pre-silicon and therefore not exhaustive). As for the high-supply-voltage region, when compared to [23], the DML MAC showed a performance gain of $\approx \times 1.11$ while reducing the energy consumption significantly by $\approx \times 1.9$. Table 9.1 also presents the measurement results for the CMOS MAC that was designed and measured in the same environment since it is a natural candidate for comparison. The results clearly illustrate the unique tradeoff of the self-adjusted MAC architecture for all modes of operation.

9.4 Conclusion

This chapter evaluated the DML technique in an FD-SOI 28 nm technology node for a very broad supply voltage operation range and aggressive (far from nominal) operating conditions. The flexibility of DML gates to operate either in the static or dynamic mode to provide E–D optimization at runtime was illustrated by silicon measurements in 28 nm FD-SOI.

The findings showed once again that the combination of DML gates operating in the dynamic mode (according to the longer logic paths), while the remainder save energy by operating in the static mode, leads to improvement in both speed and energy in terms of the actual circuit workload for a wide range of supply voltages. As a test case, an 8 × 8 bit MAC unit was fabricated and compared to an equivalent full-custom CMOS implementation. The experimental results showed that the DML MAC outperformed its CMOS counterpart in terms of speed (46%), energy (35%), and area (25%) at 0.6 V. Furthermore, when compared to state-of-the-art designs, our self-adjusted DML MAC presented unique E–D tradeoffs in all modes of operation.

References

1. D. Jacquet, F. Hasbani, P. Flatresse, R. Wilson, F. Arnaud, G. Cesana, T. Di Gilio, C. Lecocq, T. Roy, A. Chhabra, et al. A 3 ghz dual core processor arm cortex tm-a9 in 28 nm utbb fd-soi cmos with ultra-wide voltage range and energy efficiency optimization. IEEE J. Solid-State Circuits **49**(4), 812–826 (2014)
2. P. Magarshack, P. Flatresse, G. Cesana, Utbb fd-soi: A process/design symbiosis for breakthrough energy-efficiency, in *Proceedings of the Conference on Design, Automation and Test in Europe*. EDA Consortium (2013), pp. 952–957
3. D. Puschini, J. Rodas, E. Beigne, M. Altieri, S. Lesecq, Body bias usage in utbb fdsoi designs: A parametric exploration approach. Solid-State Electron. **117**, 138–145 (2016)
4. S. Jain, S. Khare, S. Yada, V. Ambili, P. Salihundam, S. Ramani, S. Muthukumar, M. Srinivasan, A. Kumar, S.K. Gb et al., A 280mv-to-1.2 v wide-operating-range ia-32 processor in 32nm cmos, in *2012 IEEE International Solid-State Circuits Conference* (IEEE, Piscataway, 2012), pp. 66–68
5. P. Flatresse, B. Giraud, J.-P. Noel, B. Pelloux-Prayer, F. Giner, D.-K. Arora, F. Arnaud, N. Planes, J. Le Coz, O. Thomas et al., Ultra-wide body-bias range ldpc decoder in 28nm utbb fdsoi technology, in *2013 IEEE International Solid-State Circuits Conference Digest of Technical Papers* (IEEE, Piscataway, 2013), pp. 424–425
6. E. Beigne, I. Miro-Panades, Y. Thonnart, L. Alacoque, P. Vivet, S. Lesecq, D. Puschini, F. Thabet, B. Tain, K. Benchehida et al., A fine grain variation-aware dynamic vdd-hopping avfs architecture on a 32nm gals mpsoc, in *2013 Proceedings of the ESSCIRC (ESSCIRC)* (IEEE, Piscataway, 2013), pp. 57–60
7. R. Taco, I. Levi, M. Lanuzza, A. Fish, Low voltage logic circuits exploiting gate level dynamic body biasing in 28 nm utbb fd-soi. Solid-State Electron. **117**, 185–192 (2016)
8. A. Kaizerman, S. Fisher, A. Fish, Subthreshold dual mode logic. IEEE Trans. Very Large Scale Integr. (VLSI) Syst. **21**(5), 979–983 (2012)
9. M. Alioto, G. Palumbo, M. Pennisi, Understanding the effect of process variations on the delay of static and domino logic. IEEE Trans. Very Large Scale Integr. (VLSI) Syst. **18**(5), 697–710 (2009)
10. G. Desoli, N. Chawla, T. Boesch, S. Singh, E. Guidetti, F. De Ambroggi, T. Majo, P. Zambotti, M. Ayodhyawasi, H. Singh et al., 14.1 a 2.9 tops/w deep convolutional neural network soc in fd-soi 28nm for intelligent embedded systems, in *2017 IEEE International Solid-State Circuits Conference (ISSCC)* (IEEE, Piscataway, 2017), pp. 238–239
11. T.T. Hoang, M. Sjalander, P. Larsson-Edefors, A high-speed, energy-efficient two-cycle multiply-accumulate (mac) architecture and its application to a double-throughput mac unit. IEEE Trans. Circ. Syst. I Regul. Pap. **57**(12), 3073–3081 (2010)
12. M.C. Wen, S.J. Wang, Y.N. Lin, Low-power parallel multiplier with column bypassing. Electron. Lett. **41**(10), 581–583 (2005)
13. I. Levi, A. Fish, Dual mode logic—design for energy efficiency and high performance. IEEE Access **1**, 258–265 (2013)
14. M. Wen, S.J. Wang, Y.N. Lin, Low power parallel multiplier with column bypassing, in *2005 IEEE International Symposium on Circuits and Systems*, vol. 2 (2005), pp. 1638–1641
15. P. Behrooz, *Computer Arithmetic: Algorithms and Hardware Designs* (Oxford University Press, Oxford, 2000), pp. 19:512,583–512,585
16. I. Levi, A. Kaizerman, A. Fish, Low voltage dual mode logic: Model analysis and parameter extraction. Microelectron. J. **44**(6), 553–560 (2013)
17. I. Levi, A. Belenky, A. Fish, Logical effort for cmos-based dual mode logic gates. IEEE Trans. Very Large Scale Integr. (VLSI) Syst. **22**(5), 1042–1053 (2013)
18. R.G. Dreslinski, M. Wieckowski, D. Blaauw, D. Sylvester, T. Mudge, Near-threshold computing: Reclaiming moore's law through energy efficient integrated circuits. Proc. IEEE **98**(2), 253–266 (2010)

19. S.S. Mahant-Shetti, P.T. Balsara, C. Lemonds, High performance low power array multiplier using temporal tiling. IEEE Trans. Very Large Scale Integr. (VLSI) Syst. **7**(1), 121–124 (1999)
20. G. de Streel, D. Bol, Impact of back gate biasing schemes on energy and robustness of ulv logic in 28nm utbb fdsoi technology, in *Proceedings of the 2013 International Symposium on Low Power Electronics and Design* (IEEE Press, Piscataway, 2013), pp. 255–260
21. A.A. Vatanjou, T. Ytterdal, S. Aunet, 28 nm utbb-fdsoi energy efficient and variation tolerant custom digital-cell library with application to a subthreshold mac block, in *2016 MIXDES-23rd International Conference Mixed Design of Integrated Circuits and Systems* (IEEE, Piscataway, 2016), pp. 105–110
22. H. Reyserhove, N. Reynders, W. Dehaene, Ultra-low voltage datapath blocks in 28nm utbb fd-soi, in *2014 IEEE Asian Solid-State Circuits Conference (A-SSCC)* (IEEE, Piscataway, 2014), pp. 49–52
23. J.P. Cerqueira, M. Seok, Temporarily fine-grained sleep technique for near-and subthreshold parallel architectures. IEEE Trans. Very Large Scale Integr. (VLSI) Syst. **25**(1), 189–197 (2016)

Chapter 10
Conclusion

In this book we presented dual mode logic, a new paradigm for digital IC, and cover multiple aspects of DML utilization in digital circuits and systems in depth. The overarching DML approach is based on DML gates that operate in two modes, each optimized for a different design metric. DML gates can trade off energy efficiency and high performance at the circuit and architecture levels. This is because DML architectures enable on-the-fly switching between operational modes at the gate, block and system levels, thus enabling significant optimization flexibility without compromising robustness. Numerous control mechanisms for DML architectures were covered in this volume.

We hope to have given the reader a thorough introduction to DML gate-level design methodology, single and multiple gate optimization, and the architectural optimization of modules and larger constructs such as arithmetic circuits. We examined several different control strategies for DML designs based on the input data and the architecture. We also demonstrated new approaches for DML integration into standard design flows in a scalable way and showed how DML can enhance technologies such as FD-SOI, thus demonstrating that DML can harness this technology to provide larger gains.

This book is intended for researchers, engineers, and graduate students. Any interested reader can find detailed responses as to how and where to use DML and what types of improvements and flexibility it can provide.

Extensive studies conducted by a range of researchers in addition to the authors make it clear that DML can enrich a whole host of fabricated designs in a variety of technologies and different architectures, supported by fully custom to standard cell-based flows. Today, the DML effort has made enormous strides forward: DML has been demonstrated in a FINFET 16 nm complex SoC, and it is also being calibrated to boost many advanced architectures within complex datapaths, processing, and arithmetic blocks. We express our deepest gratitude to all the researchers whose contributions made this book a reality.

© Springer Nature Switzerland AG 2021
I. Levi, A. Fish, *Dual Mode Logic*, https://doi.org/10.1007/978-3-030-40786-5_10

Appendix A
SA Method for the Sizing Factors of DML Inverter Chain

This appendix describes a semi-approximated (SA) approach of the LE methodology, presented in Sect. 3.3.5. The goal of SA, which is a compromise between the CA and the CS methods, is to achieve relatively high precision with reduced computational effort as compared to the CS method. Omitting all terms of the gate or drain capacitances in (3.20) may lead to increased error when calculating the delay. This error can primarily be ascribed to the first and the second terms of (3.20). Therefore, the SA approach only approximates terms starting from stage $i = 3$, denoted by an overbar. Thus, (3.20) turns into:

$$D = \sum_N D_i = t_{p0\_DML} \left(\begin{array}{c} \frac{(2s_1+1)}{3s_1}\gamma' + \frac{(s_2+1)}{2s_1} \\[2mm] + \sum_{\substack{odd\_i > 1 \\ Type\_A\ :\ 3,5,7\ldots}} \left(\frac{(2s_i+\overline{1})}{3s_i}\gamma' + \frac{(s_{i+1}+\overline{1})}{2s_i} \right) \\[2mm] + \left(\mu_{n/p}\left[\frac{(2s_2+1)}{3s_2}\gamma' + \frac{(s_3+1)}{2s_2} \right] \right) \\[2mm] + \sum_{\substack{even\_i > 2 \\ Type\_B\ :\ 4,6,.8\ldots}} \left(\mu_{n/p}\left[\frac{(2s_i+\overline{1})}{3s_i}\gamma' + \frac{(s_{i+1}+\overline{1})}{2s_i} \right] \right) \end{array} \right). \tag{A.1}$$

Differentiating (A.1) by all S_i and equating to 0 leads to the following set of N expressions:

$$\frac{dD}{dS_2} = 0 \;\rightarrow\; \frac{S_2}{S_1} = \frac{(\gamma+1+S_3)}{S_2}\mu_{n/p}$$
$$\forall\,(odd\_i > 1)\,,\ (3,5,7\ldots):\quad \frac{S_i}{S_{i-1}} = \frac{S_{i+1}}{S_i\cdot\mu_{n/p}} \tag{A.2}$$
$$\forall\,(even\_i > 2)\,,\ (4,6,8\ldots):\quad \frac{S_i}{S_{i-1}} = \frac{S_{i+1}\cdot\mu_{n/p}}{S_i}$$

© Springer Nature Switzerland AG 2021
I. Levi, A. Fish, *Dual Mode Logic*, https://doi.org/10.1007/978-3-030-40786-5

The solution to this set of equations results in the S_i sizing factors, where S_2 is the solution to the quadratic equation:

$$S_2(A_1) = \frac{\sqrt{A_1\mu_{n/p}} + \sqrt{A_1\mu_{n/p} + 4(\gamma + 1)\mu_{n/p}}}{2}, \tag{A.3}$$

and A_1 is given by:

$$A_1 = \left[\frac{S_{N+1}}{S_1} \frac{2}{1 + \sqrt{1 + 4(\gamma + 1)/A_1}}\right]^{\frac{2}{N}}. \tag{A.4}$$

Note that the calculation of A_1 requires extraction of S_{N+1} from C_{Load}. The set of equations (A.2) can be solved for any C_{Load} and any N using Matlab or a similar tool to produce one lookup table for increased user convenience/automation.

As shown, the difference between the SA and CA methods lies in the addition of the term in (A.3). Thus, in order to utilize the SA method, the sizing factors should be calculated from the F_{DML} and A_1 metrics:

$$A_1 = f_{DML} = \sqrt[N]{F_{DML}} \tag{A.5}$$

$$F_{DML} = \frac{S_{N+1}}{S_1} \frac{2}{1 + \sqrt{1 + 4(\gamma + 1)/A_1}} = f_{DML}^{\frac{N}{2}} \tag{A.6}$$

To calculate the optimal chain length N_{opt}, under a given C_{Load}, (A.5) and (A.6) are substituted in (A.1) to obtain the delay D:

$$D = t_{p0\_DML} \left(\begin{array}{l} \frac{(2s_1+1)}{3s_1}\gamma' + \frac{(s_2+1)}{2s_1} + \mu_{n/p}\left[\frac{(2s_2+1)}{3s_2}\gamma' + \frac{(s_3+1)}{2s_2}\right] + \\ + \left(\frac{(N-2)}{2}\gamma(1 + \mu_{n/p}) + (N-2)\frac{\sqrt{\mu_{n/p}}A_1^{0.5}}{2}\right) \end{array} \right) \tag{A.7}$$

Consequently, S_1–S_3 from (A.3) and Table A.1 are substituted in (A.7), in which, using (A.8), N is then differentiated and 0 equated:

$$N(A_1) = \frac{2\ln\left(\frac{2S_{N+1}}{S_1(1+\sqrt{A_1+4(1+\gamma)})}\right)}{\ln(A_1)} \tag{A.8}$$

Table A.1 Inverter chain sizing factors, S_i, of the SA method

S_1	S_2	S_3	S_4	S_5	S_{N-1}	S_N
1	$S_2(A_1)$	$\frac{A_1^{0.5}}{\sqrt{\mu_{n/p}}}S_2(A_1)$	$A_1 S_2(A_1)$	$\frac{A_1^{1.5}}{\sqrt{\mu_{n/p}}}S_2(A_1)$	$A_1^{\frac{N}{2}-1}S_2(A_1)$	$\frac{A_1^{\frac{N-1}{2}}}{\sqrt{\mu_{n/p}}}S_2(A_1)$

Finally, we get (A.9) which only contains A_1:

$$\left(\sqrt{\mu_{n/p}} \left[\left(A_1^{-0.5} + b\right) \left[\frac{1}{8} + \frac{-\frac{(1+\gamma)}{2}}{\left[\frac{2S_2(A_1)}{\sqrt{\mu_{n/p}}}\right]^2} \right] - \frac{1}{4}A_1^{-0.5} + N(A_1) \cdot \frac{A_1^{-0.5}}{4} \right] * \right.$$
$$\left. * \left[\frac{\ln(A_1)}{\frac{4b \cdot (1+\gamma)}{A_1^2(1+b)} - \frac{N(A_1)}{A_1}} \right] + \left(\frac{\gamma}{2}(1 + \mu_{n/p}) + \frac{\sqrt{\mu_{n/p}} * A_1^{0.5}}{2} \right) \right) = 0,$$

(A.9)

where, $b = (A_1 + 4(1 + \gamma))^{-0.5}$. To get the optimal number of stages N_{opt}, we further numerically solve (A.9) for A_1 and substitute A_1 in (A.8).

Index

© Springer Nature Switzerland AG 2021
I. Levi, A. Fish, *Dual Mode Logic*, https://doi.org/10.1007/978-3-030-40786-5

Printed in the United States
by Baker & Taylor Publisher Services